Fa
to Learn

the BP Texas City Refinery disaster

Disclaimer

No person should rely on the contents of this publication without first obtaining advice from a qualified professional person. This publication is sold on the terms and understanding that (1) the authors, consultants and editors are not responsible for the results of any actions taken on the basis of information in this publication, nor for any error in or omission from this publication; and (2) the publisher is not engaged in rendering legal, accounting, professional or other advice or services. The publisher, and the authors, consultants and editors, expressly disclaim all and any liability and responsibility to any person, whether a purchaser or reader of this publication or not, in respect of anything, and of the consequences of anything, done or omitted to be done by any such person in reliance, whether wholly or partially, upon the whole or any part of the contents of this publication. Without limiting the generality of the above, no author, consultant or editor shall have any responsibility for any act or omission of any other author, consultant or editor.

Failure
to Learn

CCH
a Wolters Kluwer business

the BP Texas
City Refinery
disaster

Andrew Hopkins

CCH AUSTRALIA LIMITED
GPO Box 4072, Sydney, NSW 2001

Head Office North Ryde
Phone: (02) 9857 1300 Fax: (02) 9857 1600

Customer Support
Phone: 1 300 300 224 Fax: 1 300 306 224
www.cch.com.au

Book Code: 34050A

About CCH Australia Limited

CCH Australia is a leading provider of accurate, authoritative and timely information services for professionals. Our position as the "professional's first choice" is built on the delivery of expert information that is relevant, comprehensive and easy to use.

We are a member of the Wolters Kluwer group, a leading global information services provider with a presence in more than 25 countries in Europe, North America and Asia Pacific.

CCH — *The Professional's First Choice.*

Enquiries are welcome on **1300 300 224**

National Library of Australia Cataloguing-in-Publication Data

Hopkins, Andrew, 1945–.

Failure to learn: the BP Texas City refinery disaster.

ISBN 978 1 921322 44 0 (pbk).

Includes index.

British Petroleum Company. Texas City — Explosion, 2005.
Esso Australia. Longford Plant — Explosion, 1998.
Petroleum refineries — Accidents — Texas — Texas City.
Petroleum refineries — Texas — Texas City — Safety measures.
Industrial accidents — Investigation — Texas — Texas City.
Gas industry — Accidents — Victoria — La Trobe River Valley.
Gas industry — Victoria — La Trobe River Valley — Safety measures.

363.11966553

Reprinted February 2009, Reprinted July 2009, Reprinted January 2010
Reprinted June 2010, Reprinted Sept 2010, Reprinted August 2011
Reprinted May 2012 M

Printed in Australia by McPherson's Printing Group

Foreword

On 23 March 2005, the US Chemical Safety and Hazard Investigation Board (CSB) found itself with a dilemma. The biggest explosion in 15 years had just occurred at the BP Refinery in Texas City, with multiple fatalities, missing and injured. All investigators were assigned to other important cases. The CSB management decision and will of the board chair was to put all other cases on hold and assemble a team to investigate this tragic incident in Texas City. This team would mount the most complicated and far-reaching investigation ever undertaken by the CSB since it was founded in the late 1990s.

As the team leaders returned from Texas City to Washington, they had many mechanical clues and facts to report as to what had caused this event. For months, I asked myself how this could have happened at a company like BP. I had been impressed with BP's process safety professionals and had heard presentations touting their ideas about process safety values. The mechanical and human conditions that existed at the moment of the explosion certainly explained how it happened; how, but not why.

Members of the team were already familiar with Andrew Hopkins' book about the disaster at Longford. Then, shortly into the investigation, Hopkins published *Safety, Culture and Risk*. It became apparent that the situation at BP reflected a failure to learn and transfer lessons from what had occurred at Longford years earlier.

Impressed by his book *Safety, Culture and Risk*, we invited Hopkins to the CSB to discuss the cultural conditions and the facts of the BP event. He was instrumental in teaching our staff about the links between failure to learn, or even recognise, mistakes as part of the culture of risk prevention and safety.

During speaking engagements on the importance of culture and the BP lessons learned, I have recommended Hopkins' books more than any others. Hopkins is a true teacher in the mode of all good teachers. He presents real situations and allows the reader to realise that mechanical or procedural failures are only the base of the problem. In actuality, there are a series of failures in management, budgetary priorities, and corporate values that set every stage preceding the unfolding of tragedy. Hopkins has become the pre-eminent voice in making the upward journey through the rubble of disaster to find those links with management decision-making that set the risk wheel in motion decades prior to an event.

Unfortunately, many companies, not just large multinational companies like BP, are either "risk blind" or in "risk denial" that devastating risk exists in their own operations. Maybe Hopkins' masterful teaching will remove the blinders and disbelief as he presents this story of the tragedy at BP which took 15 lives, maimed countless others, and cost BP billions of dollars and much of its reputation as an industry process safety leader. The trick here is that executive managers must read the books and reports that will help to remove the blinders and restore their sight so that they may recognise their own failure to learn.

Carolyn Merritt

Chair of the CSB at the time of
the BP Texas City Refinery inquiry

Contents

About the Author

Andrew Hopkins is Professor of Sociology at the Australian National University. He has been involved in various government occupational health and safety (OHS) reviews and completed consultancy work for major companies in the resources sector, as well as speaking regularly to audiences around the world about the causes of major accidents.

Hopkins was an expert witness at the Royal Commission into the causes of the fire at Esso's gas plant at Longford in Victoria in 1998, and in 2001 was the expert member of the Board of Inquiry into the exposure of F111 maintenance workers to toxic chemicals at Amberley Air Force base. He was a consultant to the US Chemical Safety and Hazard Investigation Board (CSB) during its investigation into the 2005 explosion at BP's Texas City Refinery and was an expert commentator in the CSB film about that disaster. In 2008, he received the European Process Safety Centre prize for extraordinary contribution to process safety in Europe. This was the first time the prize was awarded to someone outside of Europe.

As a respected OHS expert, Hopkins has authored numerous articles on the management and regulation of OHS, as well as several books which have focused on the organisational and cultural causes of major accidents. These books examine the impact that OHS law has on various industries, using interviews and examples from real life disasters. Hopkins discusses the ramification of such events and the often controversial view that effective OHS law must hold the top corporate leaders responsible when something goes seriously wrong, regardless of whether they were personally at fault.

These books, all published by CCH Australia (www.cch.com.au), include:

- *Lessons from Longford: the Esso gas plant explosion*
- *Lessons from Longford: the trial*
- *Safety, Culture and Risk*
- *Lessons from Gretley: mindful leadership and the law*
- *Failure to Learn: the BP Texas City Refinery disaster.*

Andrew Hopkins has a BSc and an MA from the Australian National University in Canberra, a PhD from the University of Connecticut, and is a Fellow of the Safety Institute of Australia.

Preface

Many readers will wonder how an Australian social scientist ended up writing a book about a disaster on the other side of the world. The explanation is that I was invited to Washington by the US Chemical Safety and Hazard Investigation Board (CSB) while it was carrying out an investigation into the Texas City explosion, in order to talk about my work on the cultural and organisational causes of major accidents. The CSB's investigation manager, Bill Hoyle, then pressed me to write a book about this incident and, in the end, he prevailed.

This book is intended for the non-technical reader. As far as possible, I have avoided using terms such as "blowdown drum" and "ISOM unit", so that readers unfamiliar with refinery technology will not be put off. I would describe this as a work of organisational sociology, about how and why organisations fail to learn, with relevance far beyond the petroleum industry. My hope, too, is that it will be of value to policymakers, in both corporate and government spheres.

I would like to thank the following people for their comments and corrections: Anthony Hopkins, Tamar Hopkins, Bill Hoyle, Kerry Jacobs, Cheryl Mackenzie, Heather McGregor, Sally Traill and Stephen Young. Attorney Brent Coon, who represented Eva Rowe, and his associate Eric Newell have been most helpful in answering questions about material on their website (www.texascityexplosion.com). Sonya Welykyj provided her usual invaluable research assistance. Deborah Powell at CCH did a meticulous editing job. Needless to say, none of these people bears any responsibility for the contents of the book.

Finally, I am profoundly grateful to Carolyn Merritt, who summoned the energy to write a foreword for this book while battling her illness.

Andrew Hopkins

Canberra, July 2008
andrew.hopkins@anu.edu.au

CCH Acknowledgments

CCH Australia Ltd wishes to thank the following team members who contributed to this publication:

Editor-in-Chief:	John Stafford
Editor:	Deborah Powell
Product Director:	Laini Bennett
Product Analyst:	Karyn Ashlin
Production Manager:	Lata Prabakaran
Production:	Ravi Kandiappan and Ziana Ellieza Bt Darus
Indexer:	Mark Southwell
Head of Marketing:	Clare Audet
Marketing Manager:	Jennifer Lim
Cover:	Trina Hayes, Feathered Edge Design
	(The cover image is computer-generated and is not from the Texas City Refinery disaster.)

Chapter 1

Introduction

"My parents were my best friends, they're all I had. My life ended that day. BP ruined my life. It ended my life. That day I had to start all over."[1]

So said Eva Rowe, who lost not one, but both of her parents in an explosion in 2005 at BP's Texas City Refinery, situated on the outskirts of Houston.[2] A total of 15 people died and nearly 200 were injured in the worst industrial disaster in the United States in more than a decade. Eva was among many hundreds who have sued BP for damages. Her case was particularly important, for reasons I shall explain shortly.

This book seeks to understand why things went so very wrong for the world's second largest petroleum company. There is, however, a prior question to be answered. Why single out this disaster for special attention? After all, there is an almost endless series of such events, each unspeakably tragic for those concerned.

Could it be that BP's failure was especially culpable? That was certainly the view of the federal regulator, the Occupational Safety and Health Administration (OSHA). It fined BP $21m, nearly twice the largest fine previously imposed by OSHA.[3,4] BP itself indulged in an orgy of blame. Initially, it sacked six of its frontline operators and supervisors. Later, a BP committee examined the culpability of the plant manager and several levels of executives above him, and it recommended that most of these people be dismissed as well. These processes of blame allocation will be explored in later chapters. And while this is a fascinating subject in itself, this is *not* what makes the BP case so deserving of attention.

What is remarkable about this case is the amount of information that has come to light on the causes of the accident; most importantly, information about the internal workings of the corporation. These workings have been laid bare, in an almost unprecedented way. When I wrote about an explosion at an Exxon gas plant at Longford in Australia, I was frustrated by the fact that the public inquiries did not raise their gaze above the level of Exxon's subsidiary, Esso

---

1 CBS, *60 minutes*.

2 Eva made an even stronger statement to a US House of Representatives committee: "BP murdered my parents." See also the media statement, 4 February 2008. Available at www.texascityexplosion.com.

3 All references to dollars are to US dollars.

4 US Department of Labor: Occupational Safety and Health Administration news release, 22 September 2005.

Australia, to examine just how Exxon's culture and organisational structure might have contributed to the event. The trail stopped dead at Australia's shoreline. In contrast, the inquiry trail after the Texas City accident jumped the Atlantic to examine the role of BP's most senior officers in London. That is one of the factors that makes this accident worth studying.

Let us consider in a little more detail the nature of the inquiries that followed the Texas City accident. First, a US accident investigation agency, the Chemical Safety and Hazard Investigation Board (CSB), launched a wide-ranging inquiry. Its report went way beyond the details of the accident to include a searching examination of BP's culture. This inquiry ranks with the Columbia space shuttle accident inquiry, from which it drew inspiration, as one of the best accident investigations carried out in recent years.

The CSB's investigation had just begun when Texas City Refinery suffered a second major fire, so large that it threatened people offsite. Weeks later, the refinery suffered a third major incident, which forced the authorities to issue an order to people in the local community to take shelter. These events so shocked the CSB that it made an urgent recommendation to the BP global board of directors to commission an independent panel to inquire into "the effectiveness of BP North America's corporate oversight of safety management systems at its refineries and its corporate safety culture". The report was to be made public. To its credit, BP responded by naming a high-powered independent panel and financing an inquiry that ended up roundly criticising the way BP's five North American refineries were being managed. The Baker Report, as the panel's report became known, and the report of the CSB have together had a profound influence on the petrochemical industry, worldwide.

In addition to these public inquiries, BP launched its own internal investigation and publicly released its report. The contrast with Exxon after the Longford accident is striking. Exxon never offered a public explanation for that accident, apart from at times blaming the frontline operators. At one point, it even argued that Longford was an accident that defied explanation![5]

Moreover, as mentioned above, BP carried out a soul-searching inquiry into the extent to which senior executives above the level of refinery manager should be held responsible for the accident. This inquiry was done with the knowledge that the report might be made public in court proceedings — which, ultimately, it was. It provides a rare glimpse into the higher realms of a major multinational corporation.

Finally, the civil actions for damages have produced a mass of useful material. In a case such as this, the US legal process allows lawyers for plaintiffs to question large numbers of people, virtually from top to bottom of a corporation, in private sessions known as depositions. A deposition can last for many hours, during which time lawyers fish for incriminating material that can be put to juries, if and

---

5 Hopkins, A, *Lessons from Longford: the trial*, Sydney, CCH Australia Limited, 2002, p 29.

when trials go ahead. Researchers also have a vital interest in this material, for it contains invaluable and otherwise unavailable insights about the way people in corporations think and behave. Unfortunately, however, these depositions are not often made public.

What is unusual about the BP matter is that many of these depositions were released, at Eva Rowe's insistence. As it had done with so many other plaintiffs, BP tried to settle Eva's claim without going to trial, for an undisclosed sum. However, she wanted the documents and depositions made public so that others might learn from them, and she sought to include this as one of the terms of the settlement. BP resisted. A battle of nerves ensued and BP finally agreed to Eva's terms only hours before the jury in her case was to be selected. The material immediately appeared on her lawyer's website for all the world to study.[6] It has been invaluable in writing this book.

These sources enable us to examine the workings of BP as a whole, and to see how BP's organisational structure, and the way it did business, contributed to the accident. The study therefore has implications far beyond the petrochemical industry. It provides insights into how large organisations in general can fail, and fail catastrophically.

The wider context

The Texas City explosion was not the only nasty surprise that BP's top leaders faced at the time. The wider context is worth outlining here because it demonstrates that BP's troubles at Texas City were part of a broader pattern.

At the end of 2005, BP's revolutionary and only partially completed deep water production platform in the Gulf of Mexico, Thunder Horse, suffered a structural collapse and tipped sideways. BP acknowledged that the root cause of the problem was insufficient engineering input, driven by a desire to cut costs.[7] Repairs cost $100m.

A few months later, in March 2006, oil was discovered leaking from a BP pipeline in Alaska. The leak was caused by pipeline corrosion that had not been repaired or even identified. Evidence of further severe corrosion was then uncovered, leading to a shutdown of Prudhoe Bay, the largest oil field in the US. These problems were widely attributed to costcutting pressures. BP's safety and environmental credentials were in tatters.

But the company's troubles also extended to its trading activities. In 2003, it was fined for manipulating the US stock market. And BP staff admitted to manipulating the North American propane market in 2004, so as to create artificial shortages from which the company could then profit.

6　Website at www.texascityexplosion.com.

7　*Houston Chronicle*, 15 June 2007.

These highly publicised problems severely dented BP's reputation and undermined its share price.[8] It looked for all the world like an organisation out of control. How could this have happened?

BP had experienced phenomenal growth since Lord John Browne of Madingly took over as CEO in 1995. A series of mergers and acquisitions saw the company grow to five times its former size, second only to ExxonMobil in the oil and gas industry. As a result, Browne was regularly chosen by his peers as Britain's most outstanding businessman. But BP's economic success was apparently achieved at the expense of its ethical, safety and environmental goals. A single-minded focus on growth had somehow undermined the company's other commitments. The Texas City accident must be seen in this wider context.

Of course, all companies seek to maximise profits and it is not very helpful to attribute BP's failures to its economic success. A central question, then, is how can companies ensure that the goal of profit does not override all other considerations? Putting this another way, how can companies *organise* themselves to ensure that proper attention is paid to their social responsibilities? The inquiries following the Texas City accident enable us to provide some tentative answers to this question.

Why lessons are not learnt

Major accident investigators frequently see their role as identifying relevant lessons so that an accident such as the one under investigation will never happen again. Moreover, relatives often express the hope that the inquiry process will ensure that no one will ever again have to experience what they have gone through. This was precisely Eva Rowe's motive in seeking the release of all of the documents assembled in the litigation process.

There is, however, as one commentator has said, a "depressing sameness" about major accidents. The causes are remarkably similar and it is apparent that companies have not learnt the lessons of earlier disasters. This is particularly evident in the Texas City case. Almost every aspect of what went wrong at Texas City had gone wrong before, either at Texas City or elsewhere. Some of these earlier failures had been extensively documented and publicised, yet BP had failed to learn from them. It exhibited a quite striking inability to learn. This book therefore departs to some extent from earlier accident analyses. It does not merely seek to identify relevant lessons; it identifies the many ways in which BP failed to learn from earlier events and it explores the reasons for BP's conspicuous failure to learn. We shall find that BP's inability to learn is attributable to the structure and functioning of the corporation as a whole.

---

8   On 19 June 2007, the *Financial Times* reported that, since the start of 2005, BP shares had underperformed the European oil and gas sector by 16%. Bloomberg reported on 18 December 2007 that BP shares had risen 15% since the explosion on 23 March 2005 versus a 72% rise for the 13-member Amex Oil Index.

The accident sequence

The accident sequence began when operators overfilled a 170-foot distillation column. These columns are common at oil refineries and gas processing plants and their basic purpose is to separate petroleum mixtures into their constituent components. If a mixture containing two liquids with different boiling points is heated, one will boil off first, thus separating the mixture into a gaseous component and a liquid component. The liquid component can then be drawn off at the bottom, and the gas at the top of the column. When operating as intended, the inflow of mixture, perhaps midway up the column, is balanced by an outflow of "purified" liquid at the bottom and gas from the top. Operators are supposed to maintain the liquid in the column at a relatively low level, but in this case, astonishingly, they filled the column almost to the top. Why they made such a serious mistake will be examined later.

As a result of this mistake, a mixture of liquid and gas flowed out of the gas line at the top of the column, travelled through emergency overflow piping, and was discharged from a tall vent which was located hundreds of feet away from the distillation column (see Figure 1.1). The discharge was described as a geyser-like eruption (see Figure 1.2). Nearly a road-tanker-load of gasoline was released in

Figure 1.1: CSB model of the accident site (the distillation column is on the far right; the vent is the left-most of the three vertical structures)

Figure 1.2: Geyser-like discharge (CSB simulation)

this way in a little under two minutes.[9] Ideally, such a vent should have a small, continuously burning flame at its top, so that any unexpected discharges are automatically ignited. This would have produced a large fire at the top of the vent and burning liquids might have rained down to the ground below, but it would not have resulted in an explosion, since the continuous flow of fuel could only have ignited as it reached the atmosphere. However, there was no such flare system (for reasons we shall examine later), and a vapour cloud accumulated at or near ground level with the potential for a massive explosion.

The vapour cloud was ignited by a vehicle that had been left in the area with its engine idling. Finally, a number of mobile offices that had been located far too close to the plant were destroyed by the explosion, killing and injuring their occupants.

Each step in this accident sequence will be scrutinised in later chapters. Each is a failure that could, and should, have been avoided.

The structure and themes of this book

After this introduction, the next four chapters follow the accident sequence described above and identify the common failing — a blindness to catastrophic

---

9   *Investigation Report: Refinery Explosion and Fire*, Washington, US Chemical Safety and Hazard Investigation Board, March 2007, p 64.

risk. Then, the precise nature of the failure to learn is pinpointed. Major accident inquiries had previously demonstrated that catastrophic risks need to be treated quite differently from other risks. BP should have learnt this, but hadn't. Why hadn't it? Chapters 8 to 11 show how BP's organisational structure and functioning impeded learning. Among the factors identified are the company's decentralised structure, the remuneration systems in use, BP's relentless cost cutting, and the lack of attention that top leaders paid to safety. Chapter 15, the conclusion, provides a diagrammatic representation of this argument.

Apart from the failure to learn, this book has several themes that are worth flagging here. One is that blame is the enemy of understanding. Most but not all major accidents are triggered by operator errors, and the initial response by companies is to blame the operators. However, operator error is better seen as a starting point for inquiry, rather than an explanation in its own right. As soon as we ask *why* operators made the mistakes they did, a whole range of factors come into view that are far more important from an accident prevention point of view. Accordingly, Chapter 2 asks a series of *why* questions about the astonishing mistake that initiated the Texas City accident, and the answers, as we shall see, lead all the way to the CEO.

This is not an argument against holding people accountable. The problem is that trying to hold employees accountable *following an accident* is likely to end in blaming them unfairly. This is as true for managers as it is for frontline employees. On the whole, it is far better to hold people accountable for inadequate performance in the normal course of events, that is, before accidents occur.

There is an exception to this rule. It is appropriate to hold the most senior people in the corporation accountable for major accidents, but this may need to be a form of accountability *without fault*. This strange-sounding idea is developed in Chapter 14.

Another central concept of this book is culture. This was also a pivotal idea for the CSB. As it said at the release of its report:

> "For the first time in its nine-year history, the CSB conducted an examination of corporate safety culture. As the science of major accident investigations has matured, analysis has gone beyond technical and system deficiencies to include an examination of organizational culture ... Effective organizational practices such as encouraging the reporting of incidents and allocating adequate resources for safe operation are required to make safety systems work successfully."[10]

---

10 US Chemical Safety and Hazard Investigation Board statement on the release of its report, 20 March 2007.

The Baker Panel followed the CSB's lead and focused its inquiry on BP's organisational culture.

This book, too, is an inquiry into BP's organisational culture and its impact on safety. But it is not explicitly structured in this way. The failure to learn is the most striking aspect of the story, and I have chosen to use this as the main focus.

There is another reason I did not use culture as the explicit organising idea, as I did in a previous book, and that is that culture is a complex and much misunderstood idea.[11] I discuss this problem in Chapter 13.

The high-reliability organisation (HRO) is an ideal to which many organisations now aspire. BP Refining was actively seeking to introduce an HRO culture. I therefore compare BP to the HRO ideal at various points, and I deal most systematically with this idea in Chapters 11 and 13.

Comparisons with the Longford explosion are also relevant. I shall argue that BP might have been expected to learn from this incident, but didn't.

It is sometimes said that the term "accident" should not be used when talking about an incident such as the Texas City explosion because it suggests that there was nothing that could have been done to prevent it. As the judge said at the conclusion of the Esso Longford trial: "What [happened] ... was no mere accident. To use the term accident denotes a lack of understanding of responsibility and a lack of understanding of cause."[12] But the word "accident" has multiple meanings, one of which is "unintended outcome". The Texas City explosion was certainly unintended and, in this sense, accidental. For this reason, I use the word freely throughout this book.

Final thoughts

The doyen of petrochemical industry disaster studies, Trevor Kletz, has said that organisations have no memory, only individuals do. But individuals eventually move on, taking their knowledge with them.[13] Organisations must therefore find ways to embed lessons from accidents into their organisational structure and functioning. This book offers some ideas on how that can be done. Ultimately, says Kletz, accident prevention depends on educating people at all levels of an organisation about previous accidents and how they occurred. Texas City is an invaluable case study for this purpose.

---

11   Hopkins, A, *Safety, culture and risk*, Sydney, CCH Australia Limited, 2005.

12   Hopkins, A, *Lessons from Longford: the trial*, Sydney, CCH Australia Limited, 2002, p 5.

13   Kletz, T, *Still going wrong!*, Amsterdam, Elsevier, 2003, p 210.

Chapter 2

Why did operators do the wrong thing?

The first step in the chain of events leading to the explosion at Texas City was an extraordinary failure by plant operators. They were supposed to keep the 170-foot high distillation column nearly empty, with only a few feet of liquid in the bottom. Instead, they filled it almost to the top (see Figure 2.1). The plant was being brought back into operation after a period of maintenance and, as operators heated the nearly full column, the mixture of liquid and gas expanded and forced its way through a pressure relief system to the atmosphere. According to one plant manager, the possibility that operators would overfill the column to such an extent had never been anticipated by anyone.[1]

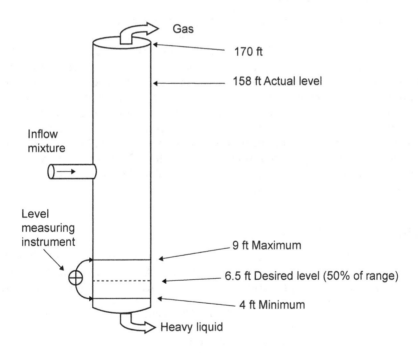

Figure 2.1: Simplified diagram of distillation column

1   Lucas, K, deposition, vol 1, 12 December 2005, p 93.

BP's initial analysis pointed to a series of violations by frontline staff. First, the procedures specifying the required liquid level in the column had been totally ignored. In particular, there was a requirement that as soon as inflow into the column began, outflow should be initiated. Instead, there was a three-hour delay in opening the outflow valve, allowing the column to fill almost to the top. Second, the rate of heating was faster than specified in the startup procedures, making loss of control more likely. Third, employees had signed documents saying that they had carried out pre-startup checks, when in fact they had not.[2] Finally, a supervisor had absented himself for some hours during the startup. The initial response by BP site management was to blame six of the workers involved and to sack them. A senior site manager who participated in the dismissal said that he had told workers and their supervisors that it was vitally important to follow procedures. He told them that "if they came across a procedure that didn't fit the activities that they were about to engage in that my commitment to them in support would be that they get the procedure right before they engage in the activity".[3] Their failure to follow procedures was, he said, "negligent". Indeed, in his view, the entire day shift and night shift had been behaving "negligently".

It is an interesting feature of human psychology that, once we have found someone to blame, the quest for explanation seems to come to an end. Accident analysts know, however, that identifying errors and violations by frontline workers is merely the starting point. We need to ask *why* the behaviour occurred and, as soon as we do, a whole range of contributory factors comes into view. If we are to have any chance of preventing further accidents, these factors must be understood and rectified. Furthermore, once we understand why the operators did what they did, it no longer seems appropriate to blame them, and punishment seems like little more than scapegoating. I shall say more about BP's attempts to hold people accountable in Chapter 12. What is of interest in this chapter is *why* the operators did as they did. What led them to behave in a way that their leadership found so unimaginable?

"Casual compliance"

BP managers had previously recognised that Texas City had what they called a culture of "casual compliance"; in other words, they recognised that employees frequently failed to comply strictly with procedures.[4] Management had therefore

---

2   *Investigation Report: Refinery Explosion and Fire* (CSB Report), Washington, US Chemical Safety and Hazard Investigation Board, March 2007, p 48.

3   Willis, K, deposition, vol 1, 14 December 2005, p 26.

4   The problem was not confined to Texas City. *The Report of the BP US Refineries Independent Safety Review Panel* (Washington, US Chemical Safety and Hazard Investigation Board, January 2007) found "instances of lack of operating discipline" at all five North American refineries (p xiii). "Operational discipline" is a term used by DuPont to describe how process safety risks must be managed.

developed a so-called "compliance delivery process" to "confirm" and "drive" compliance. This involved processes of auditing and, if necessary, discipline. The program slogans pointed in the right direction, for instance, "it's not what you *expect*, it's what you *inspect*" and "ensure we are actually doing what we say we are doing". But unfortunately the site did not have the necessary supervisory resources to carry this through and there was no attempt by management to ensure compliance with startup procedures for the distillation column. Operators were left very much to themselves and non-compliance became the norm.

There is, of course, nothing surprising about this. Organisations that put efficiency above all else often cut their supervisory staff to the point that senior managers lose the capacity to know what is really going on at the front line. This was one of the factors contributing to the Esso Longford accident. Moreover, it is a commonly heard complaint nowadays that frontline supervisors spend all of their time in front of computers and have no time to supervise. So-called high-reliability organisations (HROs), on the other hand, deploy whatever resources are necessary to ensure that operators at the workface really are complying with vital procedures.[5]

Ironically, BP inadvertently encouraged an attitude of casual compliance. The startup procedures were not updated, even though the process had evolved, and various critical events were simply not covered by the procedures.[6] In short, the procedures were at times inappropriate and workers necessarily developed their own. Despite this, "Texas City managers certified the procedures annually as up-to-date and complete".[7] They did this apparently without any attempt to verify with operators that the procedures were indeed adequate. As the Chemical Safety and Hazard Investigation Board (CSB) put it: ". . . managerial actions (or inactions) sent a strong message to operations personnel: the procedures were not strict instructions but were outdated documents to be used as guidance."[8] BP effectively acknowledged this status of procedures: "[Operators] were allowed to write 'not applicable' (N/A) for any step and continue the startup using alternative methods."[9]

There was one quite specific way in which BP contributed to this casual attitude to compliance. Prior to startup, workers had identified and reported various pieces of equipment on the column as malfunctioning.[10] These were not rectified

5   Bourrier, M, Elements for designing a self-correcting organisation: examples from nuclear plants. In Hale, A and Baram, M (eds), *Safety management: the challenge of change*, Oxford, Pergamon, 1998, pp 133–146.

6   CSB Report, pp 76–78.

7   CSB Report, p 78. See also *Fatal Accident Investigation Report: isomerization unit explosion interim report*, London, BP, 12 May 2005, p 17.

8   CSB Report, p 73.

9   CSB Report, p 77.

10   CSB Report, p 48.

prior to startup because there was insufficient time available. Furthermore, the startup was to occur even though technicians had not had the time to carry out checks on all of the instrumentation, as required by the procedures.[11] At least one supervisor initialled documents stating that the checks had been done when in fact they had not.[12] Given that earlier requests for repairs had gone unanswered, the supervisor concerned may well have formed the view that the startup would be occurring regardless of what the checks revealed and that the checks themselves were therefore unimportant.[13]

The rationale for non-compliance

The preceding discussion goes some way towards explaining why it was that the workers seemed relatively unconcerned about the written procedures. But let us consider in a little more detail why they departed from the procedures *in the particular way* that they did. The outflow valve was designed to control the liquid level in the distillation column automatically. Had it been in "automatic" mode, it would have kept the level in the column at about 6.5 feet. But the valve was deliberately set to manual, so that operators could keep the level at 9 feet or even higher. This was one of the non-compliant practices that had developed over several years.

From the point of view of the operators, there was a good reason for this practice. In their experience, the liquid level in the column could fluctuate wildly during the startup process. If the level dropped to zero while liquid was being pumped out of the bottom, this had the potential to damage equipment in the outflow line. A trip mechanism had therefore been fitted that would terminate the startup process in these circumstances. To guard against this possibility, operators thought it best to err on the side of what they thought was caution by overfilling the column.[14] The CSB commented on this as follows:

> "[This and other] deviations were not unique actions committed by an incompetent crew, but were actions operators . . . frequently took to protect unit equipment and complete the startup in a timely and efficient manner."[15]

---

11 Runfola, J, deposition, 22 May 2006, pp 13, 14.

12 CSB Report, p 49.

13 The behaviour of this individual was not just in violation of procedures, it was dishonest. But before passing judgment, it is worth reflecting that most readers will have had the experience of being confronted with a computer screen full of fine print with a box at the bottom inviting us to tick that we have read and understood certain terms. Many of us have no doubt ticked this box, dishonestly. We do so because we think it a matter of no consequence. Similarly, the supervisor concerned no doubt thought that what he was signing was a matter of no consequence.

14 CSB Report, pp 53, 54.

15 CSB Report, p 72.

The thinking of the operators is revealing. They were aware of the negative consequences of *underfilling* the column; they were quite unaware of the risks of *overfilling* the column. I shall come back later to this remarkable ignorance of the risks of overfilling the column. The point here is that the practices operators had developed reflected their understanding of the risks involved.

The Texas City operators were not unusual in this respect. Based on conversations I have had at other sites, operators commonly overfill distillation columns at startup to avoid the potentially disruptive effects of "wobbles", or fluctuations. The reason for these "wobbles" is that distillation columns are initially designed with steady state operation in mind. Startup issues are typically addressed only after the main design decisions have been made, with the result that column designs may not be optimal from a startup point of view.[16]

Why was BP unaware of this practice?

The practice of overfilling the distillation column had developed over several years. There had been 18 startups in the previous five years and, in the great majority of these startups, operators had placed the valve on manual and gone above the 9-foot level.[17] The fact that BP was unaware of such significant and sustained non-compliance calls for further comment.

There are various ways that BP might have discovered what was happening. Best practice at sites like Texas City is to define safe operating limits, or critical operating parameters, for all equipment. These might be limits of temperature, pressure or flow that operators must not exceed. Best practice also requires that equipment should be installed to monitor compliance with these limits and, where a limit is exceeded, this should generate an incident report that automatically goes to management.

Texas City had indeed established safe operating limits for some of its activities.[18] Information about instances where safe operating limits had been exceeded (exceedences) was stored centrally and was available to those who might be interested. But the incident reporting system that would have highlighted these exceedences was not operational. Much of the value of the system was therefore not realised. More importantly in the present context, no safe operating limits had ever been defined for the liquid level in the distillation column and therefore there could be no automatic alerts to management, no matter how far operators exceeded the required level.

---

16   Macnaughton, N, personal communication, 12 December 2007.

17   CSB Report, pp 73, 285.

18   CSB Report, p 100.

A second way that BP might have discovered what was going on was if senior managers had made it their business to ensure that operators were following procedures. They didn't do so, in part, they said, because in their view operators knew more about their job than the managers did.[19]

But while it may be true that operators know more about a particular job than managers do, this does not relieve managers of the responsibility of ensuring that operators are running the plant safely. Senior managers could, for instance, have sent in people with the requisite expertise to be their eyes and ears in this respect, and to report to them on whether operators were complying with procedures. Putting this another way, they might have requested compliance audits. It is, of course, difficult to audit compliance with startup procedures. Startups are sporadic and somewhat unpredictable in their timing, and those charged with auditing compliance need to be extremely flexible, as well as extremely knowledgeable, in order to catch the action and identify instances of non-compliance. But in the absence of electronic surveillance, this type of human surveillance is the only way in which managers can be sure that operators are doing as they should. At Texas City, there was no such surveillance.[20] A well-resourced process safety management department might have had the capacity to carry out this surveillance. But the process safety group was severely under-resourced, as will be discussed in Chapter 10. The end result was that Texas City managers had no way of knowing what their operators were really doing.

Why did operators overfill the column *to such an extent?*

It is one thing to fill the distillation column to the 9-foot level, or slightly above, to protect against rapid level fluctuations. It is quite another to fill the column to *158 feet*. At that level, the column could not function as intended and by any reckoning was in an extremely dangerous condition. This was the error that senior managers later said was unimaginable. Why did operators overfill the column *to such an extent*? A variety of contributing factors will be identified in the following sections.

Why was it *possible* to make this mistake?

In exploring why operators overfilled the distillation column to such an extent, we should begin by asking why it was physically *possible* to make this mistake. It would have been a relatively simple matter to install a cut-out device that prevented the operators from overfilling the column to the extent that they did. But this would have required additional expenditure, which BP was seeking to

---

19   Willis, op cit, p 36; Hawkins, R, deposition, 16 February 2006, p 14.

20   *Fatal Accident Investigation Report* (FAIR), London, BP, 9 December 2005, p 142.

avoid. The company view was that, if operators followed procedures, there should be no need for such a backup safety mechanism. Management's considered position *prior* to the accident was as follows:

> "In the face of increasing expectations and costly regulations we are choosing to rely wherever possible on more people-dependent and operational controls rather than preferentially opting for new hardware. This strategy [will place] greater demands on work processes and staff to operate within the shrinking margin for error."[21]

In adopting this policy, BP was ignoring an abundance of evidence from other accident investigations that systematic deviation from stated procedures is the norm, rather than the exception. Indeed, the evidence is that operators sometimes feel obliged to violate stated procedures in order to maintain production. That being so, the greater the reliance that management places on operators following procedures, the greater the need for close supervision to ensure that they are indeed working "within the shrinking margin for error". However, as already noted, Texas City did not have the necessary supervisory resources to ensure compliance. BP's decision to place primary reliance on procedures, without recognising that additional resources might be necessary to ensure compliance with those procedures, must be counted as one of the causes of this accident.

The decision to rely on procedures in this way is contrary to the well-known hierarchy of controls for dealing with hazards. There are various versions of the hierarchy, but one simplified version identifies the following controls, in descending order of effectiveness: elimination, substitution, engineering controls, administrative controls, and personal protective equipment (PPE). Let me spell this out in a little more detail.

At the top of the hierarchy, the ideal way to deal with a hazard is to eliminate it completely, preferably at the design stage. Failing that, it may be possible to substitute a less hazardous substance, process or piece of machinery. If the hazard cannot be eliminated or reduced, ways must be found to protect against it. There may be engineering solutions, such as automatic cut-out devices. Failing that, there may be administrative controls, that is, procedures designed to minimise the risk, and finally, if all else fails, workers can be equipped with PPE, such as helmets and fire protection suits. It is important to stress that these last three are in decreasing order of effectiveness. Engineering solutions are reasonably reliable (though, of course, they can fail if they are not properly maintained or if they can be deliberately disabled); procedures are only effective if complied with, which requires supervisory resources; and PPE can be uncomfortable, inappropriate, troublesome to maintain, and even counter-

---

21   CSB Report, p 89.

productive.[22] It can be seen, then, that in choosing to protect against disaster by relying on procedures, rather than new hardware, BP was opting for the less effective of the two strategies.

Tragically, there is a long history in the petroleum industry of accidental discharge incidents caused by the overfilling of distillation columns.[23] Various authorities have recommended that automatic shutdown devices should be installed on such columns to prevent overfilling. BP's decision to rely on procedures rather than hardware flew in the face of this advice.

In summary, in seeking to understand why the operators did what they did, we should not lose sight of the fact that it was physically *possible* to fill the column to the top. Had a cut-out device been in operation, the accident could not have happened. In this sense, the absence of such a device was one of the causes of the accident.[24]

Inadequate instrumentation

A second factor that contributed to the *extent* of the mistake made by the operators was that they were quite unaware of how full the distillation column had become. Instruments on the column were designed to indicate where the liquid level was in the range from 4 to 9 feet. Once the level went above this upper limit, there was no instrumentation to tell operators how full the column was. Above the 9-foot level, they were operating in the dark!

This is a significant instrumentation deficiency. Arguably, the design of the column was substandard in this respect. Distillation columns usually have the capacity to measure liquid levels over a much greater range.

This is not the first time that such a deficiency has led to problems. At Longford, operators were accustomed to running the plant for hours above the maximum allowable level, without any instrumentation to tell them just how far above the limit they were. This practice initiated the accident sequence at Longford, just as it did at Texas City.

But this was not the most serious problem with the instrumentation. If the level measuring instrument had been indicating that the liquid had reached the maximum measurable level of 9 feet, operators might have been more alert to the possibility that the true level was beyond the range of their instrument. However, the instrument showed that the level was slowly declining in the hours before the accident, from just under 9 feet to just under 8 feet.[25] As far as the control room operator was concerned, such a fall in level was plausible, given the

---

22 *Report of the F111 Deseal/Reseal Board of Inquiry*, Canberra, Royal Australian Air Force, 2 July 2001, pp 7-2, 7-3.

23 CSB Report, p 105.

24 The word "cause" is to be understood here as meaning "necessary condition".

25 CSB Report, pp 56, 57.

fluctuations to which the startup process was liable, and he therefore had no reason to question the accuracy of the instrument. Accordingly, his understanding was that the level was slowly returning to the midpoint of the measurable range, 6.5 feet, when in fact it was still rising.[26] This was a crucial instrument failure. There is a tragic irony here: this very instrument was one that had earlier been reported as malfunctioning but had not been fixed.[27]

In many of the previously documented accidents involving overfilling distillation columns, operators had been misled by faulty level measuring instruments.[28] Texas City management was apparently unaware of this history.[29]

Interestingly, there were two alarms designed to warn operators that the column was overfilling. The first was set at a little under 8 feet. Since the operators were intending to fill the column to 9 feet or more, they ignored this alarm.[30] The second and slightly more urgent alarm was set at the 8-foot level.[31] This alarm was one of the pieces of column equipment that was not working at the time. BP's position was that the operators should have stopped the process as soon as they realised that this alarm was not working.[32] But since the operators were intending to fill the column above that level, whether or not it was functioning was beside the point. What is really striking here is that, given the way the operators were doing their job, these alarms were essentially irrelevant. Again, the parallel with Longford is eerie. In that case, operators routinely operated the plant in alarm mode, that is, beyond the point where alarms had been triggered, in order to maximise production. The alarms were irrelevant to them, just as they were to the Texas City operators.

There was one other way that operators might have checked on the level. A sight glass was positioned on the column so that the liquid level could be directly seen. However, there was a build-up of residue on the glass and it had been effectively useless for years.[33] Requests that the glass be cleaned during maintenance periods had gone unheeded.[34]

---

26   CSB Report, pp 81, 82.

27   It can be argued that, had the column not been overfilled in the first place, the instrument would not have malfunctioned in this way (FAIR, pp 54, 58), but the fact remains that the instrument failed to indicate the true level and thus misled the operator. The nature of this failure is analysed in the CSB Report at pp 319–323.

28   CSB Report, p 81.

29   A column overfill incident caused by instrument malfunction occurred at another BP North American refinery nine months after the Texas City accident (CSB Report, p 107). It seems that management at this site was unaware of what had happened at Texas City or, at least, had been unable to learn from it.

30   CSB Report, pp 81, 82.

31   CSB Report, p 82.

32   Willis, op cit, p 28.

33   CSB Report, p 52.

34   CSB Report, p 48.

Finally, there were instruments that measured what was going into the column and what was coming out. Had the attention of the operator on duty that day been drawn to the imbalance in these two measurements, he might have realised that the column was filling up. The control room displays did not, however, highlight this imbalance.[35] Precisely this problem was found to have contributed to the overfilling of a distillation column and consequent explosion at the Milford Haven Refinery in Wales in 1994.[36] The United Kingdom authorities recommended at the time that control displays should be modified to highlight this information. This had not been done at Texas City.

In summary, the instrumentation available to operators was seriously lacking — even worse, misleading. Management had not provided operators with the ability to monitor effectively how full the column was, and the result was they lacked any understanding of what was happening. High-reliability organisations stress the need for "situational awareness" among their frontline operators. The operators at Texas City were entirely unaware of the situation they were in.

Why did *this* startup go so wrong?

Operators had overfilled the distillation column at previous startups. They did so by opening the inflow valve into the column but leaving the outflow valve closed for a period. The average delay in opening the outflow valve was 15 minutes and the longest previous delay was 46 minutes.[37] On the occasion of the accident, it was three hours before the outflow valve was opened. During these three hours, the column filled almost to the top.[38]

It is obvious that prolonged inflow without any outflow must in the end cause trouble. This same situation had arisen at Longford and counsel for Esso had cruelly remarked that "a child [knows] ... how to stop the level of water increasing in his bath. He knows to either turn off the tap or pull the plug". Such a retrospective comment would be just as unfair to the operators at Texas City as it was to the operators at Longford. But it does provoke the following questions: why did the operators delay opening the outflow valve for so long, and what was it about the circumstances on the day of the accident that was different from previous startups?

There were at least two distinctive features of this startup. First, the control room operator believed that he had been instructed not to open the heavy liquid outflow valve because the storage tanks where the liquid would be held were full.[39] There is some debate about what he was actually told, but communication

---

35   CSB Report, p 83.

36   CSB Report, p 106.

37   FAIR, p 58.

38   Ironically, opening the outflow valve contributed to the problem because it increased the rate of heating of the inflow.

39   CSB Report, p 80.

on this matter was "rushed and vague" and the fact is he understood that he had been told to leave the valve closed.[40]

The miscommunication problem was actually far worse than this. At a management meeting held on the morning the startup was scheduled to take place, the decision was made *not* to proceed, precisely because the storage tanks that received the heavy liquid were full.[41] However, operators were not told of this decision and went ahead with the startup, as originally planned. The result was that the startup was initiated without any authorisation — indeed, contrary to management intentions![42] A more extraordinary communication failure is hard to imagine. This was a site in which operational discipline was sadly lacking and where management's ability to manage was seriously deficient.

A second unusual feature of this startup is that it occurred over two shifts.[43] Previous startups had been accomplished on a single shift. Shift handover is a time when communication failures can easily occur, particularly when plant is experiencing abnormal conditions. A communication failure at shift handover was one of the precipitating factors of the Longford accident. It was likewise implicated in the Texas City accident.

Research has shown that shift handover is most effective when communication is both written and oral.[44] At Texas City, the communication in the operators' log book was brief and uninformative, and there was no face-to-face contact between the incoming and outgoing operators and therefore no oral communication at all about what the first shift had been doing. A communication failure in these circumstances was almost predictable.

The first shift had already filled the distillation column above the 9-foot level, in addition to filling to capacity the outflow pipes and pumps. However, they had not started heating the column; that had been left to the incoming shift.

The new shift did not realise the extent to which the column and all of the associated pipework were already "packed" with liquid, to use the industry jargon. Indeed, the incoming control room operator assumed that the associated pipework had *not* been filled and that further input was required.[45] Had he understood that the earlier operator had already completely filled the system to the required level in readiness for heating, he would probably have behaved differently and the accident would not have happened.[46] In short, inadequate communication between shifts was one of the causes of this accident.

---

40   CSB Report, p 80.

41   CSB Report, pp 53, 79 fn 68.

42   CSB Report, p 52 fn 39.

43   FAIR, p 81.

44   CSB Report, pp 286, 287.

45   CSB Report, p 51.

46   The CSB Report does not draw this conclusion explicitly in its text, but it is an implication of its fault tree analysis (pp 229, 232).

The lack of awareness of what might go wrong

Another factor that contributed to the long delay in opening the outlet valve was the lack of awareness by operators of the *dangers* of overfilling the distillation column. This was due in part to inadequacies in the training they had received.[47] Much of the training was on the job and concerned with routine activity. In addition, there was some computer-based training which dealt mainly with factual material, such as which alarm corresponded to which piece of equipment. BP's computer-based training had been introduced as an alternative to classroom training as a way of minimising cost, but without much attention to its educational value.[48] It did not provide in-depth understanding of the process, or what might go wrong, or why certain alarms or procedures might be critical.

In particular, the training did not deal with the hazards of overfilling distillation columns. Given that this is such a common type of accident in the industry, it would clearly be desirable to provide operators with some of this history. Studying past accidents and the lessons that emerge from them is one of the most effective ways of deepening understanding of what can go wrong. Had the operators studied the Milford Haven accident, for example, in which a column overfill resulted in an explosion that destroyed a large part of the plant, they would have been far more sensitive to the dangers that they faced.

Most importantly, there was no training on how to handle abnormal situations. On the day of the accident, operators faced abnormal pressure fluctuations that they did not understand and which they were not equipped to handle. One of the best ways for operators to learn how to handle abnormal situations is to place them in simulated control rooms and present them with a variety of problems to solve. Cockpit simulators are now standard in the airline industry and pilots spend time dealing with simulated engine failures and the like. This is not yet routine for control room operators in the petroleum industry. The training department at Texas City had pressed management to invest in simulators, but the request had always been rejected on cost grounds.[49]

Had the operators been following procedures more closely, their lack of understanding of the risks that they faced might not have been an issue. But, given that the plant was being operated with scant regard to formal procedures (indeed, according to an informal set of procedures that had been developed on the job), sooner or later the operators' inadequate training was bound to prove disastrous.

---

47 CSB Report, p 95.

48 CSB Report, p 98.

49 CSB Report, pp 96–98.

Why were operators not more alert to the warning signs?

In the hours preceding the accident, there were indications that something was wrong. Most dramatically, there were several "pressure spikes" (occasions of rapid and dangerous pressure increase). The strategy of the operators was to deal with the symptom by opening a bypass valve to relieve pressure, without ever asking why the excess pressure had occurred. The cause of the problem was, of course, that the outlet valve had remained closed, allowing the distillation column to fill almost to the top, but the operators never made this connection.

Part of the reason for this, as noted above, was that their training had not included any discussion of the consequences of overfilling.[50] As a result, they were not well equipped to understand what was happening. Be that as it may, one might have expected that the operators would at some stage recognise that they had grossly overfilled the column. How *could* they have remained so totally unaware of what had happened? Their failure is so remarkable as to require some further explanation.

The CSB provides this explanation: the problem-solving ability of the operators was degraded by fatigue.[51] At the time of the accident, the day control room operator had been working 12-hour shifts, seven days a week for 29 consecutive days! Others had worked for even longer periods without a break. This control room operator reported that he routinely got only five or six hours of sleep per night, which amounted to a sleep deficit of approximately 1.5 hours per night. Sleep research has demonstrated that nightly sleep deficits cumulate, that is, the more consecutive nights of deficit, the greater the sleep debt, and that this sleep debt affects judgment. According to the CSB, the operators' focus on the symptoms and the failure to connect the symptoms with their cause was an example of "cognitive tunnel vision — a typical performance effect of fatigue".[52]

There is good experimental evidence that fatigue reduces performance, and there is good statistical evidence that fatigue causes accidents. For instance, there are data showing that accident rates increase markedly in the last four hours of a 12-hour shift.[53] However, it is difficult to show conclusively for any particular accident that fatigue was a crucial factor, in the sense that, had the individual not been fatigued, the right judgments would have been made and the accident would not have happened. Nevertheless, the considered view of the CSB is that

---

50 CSB Report, p 95.

51 CSB Report, pp 89–93.

52 CSB Report, p 92.

53 Folkard, S, *Report to Civil Aviation Authority on Work Hours of Aircraft Maintenance Personnel*, Swansea, University of Wales, 2003. See more generally, Hartley, L, Sully, M and Gilroy, P, Development of proposed code of practice governing working hours in WA, *J Occup Health Safety — Aust NZ* 2005, 21(4): 351–368.

"by degrading judgment and causing cognitive fixation, fatigue likely contributed to overfilling of the . . . tower".[54]

The fact that operators were working 12-hour shifts on consecutive days for a month or more was not some kind of aberration. It was BP policy.[55] During maintenance shutdowns, operators were expected to work 12 hours a day, seven days a week. This was the norm, although individuals could apply for exemptions from this schedule on a case-by-case basis. Arguably, BP's lack of any fatigue prevention policy contributed to the accident.

Major hazard industries need to be especially sensitive to this issue. Nor is it sufficient to rely on individual self-assessment. As the CSB rightly observes, individuals may be unaware or unwilling to admit that they are fatigued. Shift systems must be carefully designed to minimise the risk of fatigue. In one air traffic control organisation that I have studied, shift length for controllers is designed with fatigue in mind and varies from eight hours for various day shifts down to six hours for the midnight to dawn shift. This is the kind of sensitivity that is required if fatigue is to be effectively managed.

Inadequate staffing of the control room

The control room operator responsible for the startup of the distillation column was responsible for a total of three different processing units at Texas City. BP's own estimates were that, when all three units were running in steady state condition, this was already a full-time job.[56] On the morning of the accident, the operator was also managing a startup. Given the extra manual work and additional critical thinking that is required at such times, this was considerably more than a full-time load for one person. On several previous occasions, internal BP analyses had drawn attention to the need for two operators at these times, and workers themselves had campaigned for improved staffing.[57] But Texas City policy was to staff the plant at a "lean" level, that is, with no allowance for additional workload during startups or emergencies, so as to cut costs.[58]

This is a matter about which there was considerable controversy, so it is necessary to be crystal clear at this point.[59] There is no doubt that the control room operator was busy on the morning of the accident, but BP argued, with some justification, that an additional operator would not necessarily have made a difference to the decisions that were made that morning. In other words, it

---

54   CSB Report, p 94.

55   CSB Report, p 93.

56   CSB Report, p 86.

57   CSB Report, p 291.

58   CSB Report, pp 88, 89.

59   Broadribb, M, deposition, 15 February 2006, p 35; FAIR, p 152.

cannot be said with any certainty that BP's one-operator policy was a cause of this particular accident.

But this does *not* amount to a vindication of the one-operator policy. BP's own analyses had shown that two operators were necessary at certain times to handle the workload safely. To ignore its own internal advice and to opt for a policy of "lean" staffing, especially at times of startup, suggests that BP was suffering from a form of organisational risk blindness.

High-reliability organisations are acutely sensitive to the workloads of their frontline staff. They are willing to deploy more staff than is necessary in the normal course of events so that, when abnormal situations arise, suitable staff will be on hand to analyse and assist. Such staff may be seen as redundant during times of normal operation, but this is the price HROs are willing to pay to ensure that their expertise is available when required. Texas City aspired to be an HRO, but in this matter it fell far short of its goal.

The risks of startup

The great majority of aircraft accidents occur during take-off and landing. En route flight is very low risk by comparison. No matter how long the flight, it is the short periods at the beginning and end of the flight that make by far the greatest contribution to the total risk to which passengers are exposed. The difference is so marked that accident rates for airlines are often defined as accidents or fatalities per take-off and landing, rather than per hour or per mile of flight.

The situation is comparable for refineries. There is good evidence that startup is much more risky than steady state operation.[60] Some of BP's own policy documents draw attention to the greater risks at such times. In particular, an earlier analysis of BP's North American sites concluded that incidents were 10 times more likely to occur during startup than during normal operation. A chemical industry document states that, even though startup represents only a small portion of the operating life of a plant, process safety incidents occur five times more often during startup. The implication of this last statement is worth highlighting. If we hypothesise that a particular plant may be in startup mode 5% of the time, a quick calculation reveals that the risk per hour is approximately a hundred times greater during startup than during normal operation. If it is in startup mode only 1% of the time, the risk is approximately 500 times greater, and so on. Once this is understood, it is clear that organisations should take special care and devote whatever additional resources may be necessary to ensure that this additional risk is properly managed. This would certainly have required a two-man control room operator team at Texas City, but it would also have required BP to pay special attention to what operators were doing at times of startup and to monitor their behaviour both electronically and by direct

---

60   CSB Report, p 44.

observation. Had such additional resources been forthcoming, the operators could not have developed the practices that ultimately led to disaster.

BP's failure to recognise startup as a time of increased risk introduces the theme of organisational risk blindness that will be developed in later chapters. Its significance here is that it helps account for the lack of risk awareness that was manifested by the operators on the day in question. Their behaviour was consistent with what was being modelled for them by the organisation as a whole: a risk-blind organisation almost predictably generates risk-blind individuals. Conversely, a risk-aware organisation would have done its utmost to instill risk awareness in its employees, for instance, by teaching them about what had happened at other sites around the world.

Conclusion

This chapter demonstrates just how useful it is to ask *why* the operators behaved as they did. It is true that, had they complied with the procedures, the accident would not have happened. But an explanation in terms of violations by frontline workers is entirely threadbare in comparison to the rich explanatory fabric that the preceding analysis has woven. This more detailed analysis is vital for accident prevention purposes. Moreover, once this deeper understanding has been gained, the impulse to blame the operators dissipates.

Several of the explanatory factors identified in this chapter *themselves* require explanation. Why was Texas City so wracked by cost cutting? Why had the site failed to learn from previous accidents around the world? These and other questions will be pursued in later chapters.

Chapter 3
Vapour cloud

As a result of the mistake made by operators, a vast quantity of petroleum, almost a truckload, forced its way through the emergency pressure release system and escaped to the atmosphere through a tall vent (119 feet high).[1] Witnesses described a geyser-like eruption from the top of the vent. Had this material been lighter than air, it might have dispersed without igniting. But some of it was liquid and some was heavier-than-air vapour, so the mixture fell to the ground, where a massive cloud of petrol vapour accumulated. Texas City was then only moments from disaster.

It was not inevitable that a vapour cloud accumulate in this way. The first line of defence, of course, is to ensure that there is never any "loss of containment", to use the industry jargon. The mistake by operators at Texas City breached this line of defence. The second line of defence is to ensure that, if there is a loss of containment, combustible materials are ignited *as they are released*, so that they cannot accumulate and subsequently explode. This can be done using a flare, that is, a system that ignites combustible material at the top of a tower.

The discharge from the vent on the day of the accident took place over a period of 107 seconds, that is, a little under two minutes. Had the discharge occurred through a flare, the material would have been progressively ignited over this period, at the top of the flare tower. This would have avoided any explosion. Of course, the fire would have been intense, and burning liquids would have cascaded down from the top of the tower, threatening the destruction of anything at the base. The industry is well aware of this possibility and flares are therefore located in flare yards that are fenced off, so that everything and everyone is kept at a distance. At Texas City, there were indeed flares located in flare yards. It can reasonably be assumed that, had the escaping material on the day of the accident been routed through a flare in a flare yard, the consequences of the accident would not have been nearly as catastrophic and most probably no one would have been killed.

So the question is: why was the process unit concerned connected merely to a vent and not to a flare?

At the time the vent was built in the 1950s, open vents were assumed to be safe.[2] With the passage of time, industry standards improved and, as early as 1986, Texas City was operating under a corporate standard set by its then owner,

---

1 *Investigation Report: Refinery Explosion and Fire* (CSB Report), Washington, US Chemical Safety and Hazard Investigation Board, March 2007, p 38.

2 CSB Report, p 38.

Amoco, which specified that *"new* ... [vents] that discharge directly to the atmosphere are not permitted".[3] In 1991, Amoco's planning department proposed a strategy for eliminating *existing* open vents but this did not go ahead because it was judged that government regulations were unlikely to require this in the foreseeable future.[4] This reliance on government regulations to ensure progress on safety is something that I shall return to shortly. But the end result was that Texas City operated under a corporate standard that prohibited the building of new vents but did not require existing vents to be replaced with flares.

This was an uneasy situation. In 1992, the United States regulatory agency, OSHA (Occupational Safety and Health Administration), found that an open vent at Texas City was unsafe and needed to be replaced, but in a settlement with Amoco, it agreed not to pursue the issue.[5]

Even more disturbingly, in 1997, the vent system in question was completely replaced with similar equipment. This had been an obvious opportunity to replace the vent with a flare, but the opportunity was not taken. The justification appears to have been that this was not a "new" structure, merely a replacement and, as such, the corporate policy did not apply. There is clearly room for debate about whether something that is rebuilt from the ground up is new, or merely a renewed version of the old, as one manager speculated.[6] The best that can be said about this is that it was a missed opportunity to reduce risk. BP has a policy of continuous risk reduction.[7] It is not clear whether Amoco did at the time of the rebuilding; if so, it apparently had no impact at Texas City.

Risk management versus rule compliance

The prolonged failure to upgrade the vent to a flare highlights a limitation of the risk management approach to major hazards. This is an important point that needs to be carefully developed.

In jurisdictions such as the United Kingdom and Australia, the overarching legislation requires that risks be reduced "as far as reasonably practicable". In the US, on the other hand, the overarching legislation is not explicitly based on any concept of risk reduction; it requires employers to provide a workplace that is "free from recognized hazards that are causing or are likely to cause death or

---

3   Maclean, C, deposition, 12 September 2006, p 53.

4   CSB Report, p 114.

5   CSB Report, p 24.

6   Sorrels, S, deposition, 8 June 2006, p 37. The question is one of identity and difference. One is reminded of the woodcutter who spoke lovingly about his old axe: "Mind you, I've had to replace the head a couple of times and I've been through several handles, but this old axe has served me well."

7   *The Report of the BP US Refineries Independent Safety Review Panel* (Baker Report), Washington, US Chemical Safety and Hazard Investigation Board, January 2007, pp 167–169.

serious physical harm".[8] Nor does the idea of risk appear in the subordinate process safety management standard that governs major hazard sites like refineries. Instead, the standard specifies in some detail the safety management *processes* that must be undertaken. There is an unavoidable ambiguity here: we are talking both of management *processes* and *process* industries, that is, industries that process chemicals. So to repeat, the process safety standard specifies management processes. It does not specify technical standards that must be complied with; nor does it talk about risk assessment. But when we look at the guidance provided on how to *interpret* standards, it finally becomes clear that the decisions about how to comply with the standard may ultimately involve some form of risk assessment, involving judgments about likelihood and consequences.[9] It turns out, then, that in the US, just as in jurisdictions such as the UK, the regulation of major hazard sites such as refineries requires judgments about risk, rather than compliance with prescriptive technical rules.

There is a good reason for this, namely, that it is impossible to devise a set of technical rules that cover every situation. The law therefore requires companies to manage risk themselves, although it specifies in considerable detail how this is to be done.

The net result is that there was no specific regulatory requirement that Texas City replace vents with flares; all that was required was that it manage the risk. So it was that OSHA's 1992 finding that an open vent at Texas City was in violation of the law was not a finding that it had violated any clear-cut rule, but merely that its system of discharge was not "safe" — a far from precise term. The exact words of the finding were that "the relief valve discharge lines . . . [did not] lead to a *safe* place of discharge" (emphasis added).[10] Clearly, there was room for dispute about how safe the system was and, following OSHA's initial finding, Amoco was apparently able to persuade the regulator that the design was consistent with industry standards and therefore that the risk could be considered adequately controlled.[11] For this reason, OSHA ultimately withdrew

8   *Occupational Safety and Health Act of 1970* (US), section 5. It is possible to read this general duty provision of the Act as equivalent to a requirement that "intolerable" risks must be eliminated.

9   Occupational Safety and Health Administration Standards Interpretations, *Documentation of methods used to comply with the qualitative evaluations of a range of possible safety/health effects of "failure of controls" requirement of the PSM standard*, standard number 1910.119, 1910.119(e), 1910.119(e)(3), 1 February 2005.

10 OSHA Citation and Notification of Penalty, reporting ID 0626700, issuance date 30 March 1992.

11 CSB Report, p 24 fn 10. The concept of "industry standard" can be given quite a precise meaning here. It was specified in the American Petroleum Institute (API) Recommended Practice No 521, *Guide for pressure relieving and depressurising systems*. This guideline accepted that depressuring might be via either a vent or a flare (CSB Report, p 118). The CSB argued that this was a defect in the standard and made an urgent recommendation to the API that it revise its standard to prohibit the release of flammable material to the atmosphere.

its objection as part of a settlement agreement.[12] Had a flare been mandated by government regulations or standards, Texas City would no doubt have complied; but since process safety management was ultimately a matter of risk management rather than rule compliance, Texas City was able to avoid the expense of implementing best practice.[13]

An important implication that can be drawn from this analysis is that, where it is clear what best practice is — and this will not always be the case — governments should consider formulating this practice as a specific requirement that can be imposed on industry. Had such a requirement been in place, there would have been no room for Amoco to argue with OSHA as to whether a vent was safe; it would have been required by law and that would have been the end of the matter.

One can draw an analogy with road safety here. Generally speaking, the faster we drive, the greater the risk. So how fast should we drive? The risk management approach would leave it to every individual to decide, taking account of the particular road conditions. Public policy does not, however, accept this approach, fearing that individuals will not make sound risk judgments. Instead, it imposes blanket speed limits which are inevitably arbitrary and often ill-matched to the particular road conditions. Nevertheless, the evidence is that enforcing these speed limits reduces death and injury.[14]

Arguments by analogy are never conclusive and one can often distinguish the analogous situation from the situation under discussion in ways that undermine the value of the analogy. Critics may well argue, therefore, that for one reason or another the road safety analogy is not a good one. There is, however, a comparison much more immediate than the road safety example that leaves far less room for argument.

---

12   In 2005, after the accident, OSHA cited BP for the same violation and this time BP accepted it and paid the $70,000 penalty. The citation stated: "In order to abate this violation, the employer must ensure flammable hydrocarbons are not released near employees exposing them to potential burns and death by such methods as the utilization of a flare in a remote location." Notice that what is being required here is that "flammable hydrocarbons are not released near employees", not necessarily that BP install a flare.

13   The CSB Report (p 121) notes that "published [OSHA] guidelines call for inspections to ensure that 'destruct systems such as flares are in place and operating'". It might be inferred from this that flares were indeed mandatory. However, the guidelines are for OSHA inspections and the reference to flares is provided only as an example of what inspectors might look for (see OSHA CPL 2-2.45, 1910.119(e), III, B, 1). Furthermore, the guideline at this point draws its authority from Process Safety Management Standard 119(e)(1), which makes no specific reference to flares. The guidelines cannot therefore be interpreted as specifically mandating the use of flares.

14   Yannis, G, Papadimitriou, E and Antoniou, C, Impact of enforcement on traffic accidents and fatalities, *Safety Science* 2008, 46(5): 738–750.

Environmental compliance

In 2002, Texas City seriously considered replacing the vent with a flare. It did so for environmental reasons, not safety reasons. It had been under pressure from the Environmental Protection Agency (EPA) in relation to benzene emissions from the site. Benzene is a carcinogen and the greater an individual's exposure to this substance, the greater the risk of cancer. The EPA did not, however, leave it to companies to manage this risk themselves. By law, it could create national emission standards for hazardous air pollutants, and it had declared such a standard for benzene that restricted Texas City to two "megagrams" per year. In this way, the issue of managing the risk of benzene emissions had been converted into a rule that Texas City was required to comply with. In short, risk management had been converted into rule compliance.

Texas City managers were concerned that benzene leakages might exceed this limit and they considered various ways of reducing this risk. As one said:

> ". . . we had significant concern that our operations could put us in a non-compliant position. So the project was advanced very quickly to assure us that any points of non-compliance in our operation . . . were being addressed."[15]

One way of reducing the possibility of benzene emissions would have been to convert into a flare the same vent that later contributed to the accident. In the end, Texas City found other ways of dealing with the problem that did not involve the expense of replacing the vent with a flare. Interestingly, managers were aware that there was a strong likelihood that tighter environmental rules would require them to replace the vent with a flare within the next five years.[16] But the decision was to hold off on this course of action until forced to do so by environmental regulations.

This story is instructive. It shows that where Texas City managers were confronted with a clear rule, they were prepared to go to considerable lengths to ensure compliance. This was part of the "licence to operate", as one manager said.[17] He was aware that failure to comply with the terms of the licence might have significant consequences, as he made clear during his deposition:

> Q: ". . . and part of the sense of urgency was that otherwise the EPA could invoke some rather serious sanction on BP and some of its officers, could they not?"

---

15   Wundrow, W, deposition, vol 2, 29 August 2006, p 39.

16   Hale, R, deposition, 1 June 2006, p 58.

17   Willis, K, deposition, vol 1, 14 December 2005, p 63. Of course, there was still an element of risk management involved, the risk being that the site might exceed the two-megagram limit. A lawyer made the point quite provocatively when he asked a former Texas City manager: "You are risk managing your permit [to operate] rather than risk managing your employees?" The manager took exception to this way of putting it: "I don't agree with that . . . the project was about compliance."

A: "I don't know exactly what EPA's response would have been. What I know clearly is that they had an expectation that we address this urgently and, as quickly as possible, bring ourselves back into compliance."

Q: "Okay. Did the EPA have the ability to shut the facility down if they believed that you were not doing what you needed to do to get back into compliance?"

A: "I would suspect they would have some recourse to that . . ."

Q: "Okay. Do you know that they had the authority to put people in jail if they felt that they were not in compliance and not doing enough?"

A: "Well, I have read some cases where there was certainly a . . . possibility of . . . jail sentences."

Q: ". . . that . . . creates a sense of urgency . . . Did you have a sense of urgency to fix it?"

A: "As I said, yes, it was a very urgent thing for me and for the site. Not that I felt personally threatened about going to jail. I would not be honest to say that was true, but I felt very much that being in compliance was one of operating dictates that we wanted to live by and we were not in compliance."[18]

This passage gives a sense of how important it was for managers at Texas City to be in compliance with environmental requirements. Their sense of urgency stemmed from the close scrutiny that they were under from the EPA, and beyond that from a recognition that there could be serious, if unspecified, consequences if they failed to comply.

There is an important observation that can be made here. It concerns the contrast between the attitude of Texas City managers towards compliance with EPA rules and the attitude of Texas City workers towards compliance with procedures. For the managers, compliance was a matter of urgency; for the workers, compliance was optional. Why the difference? The answer is that managers expected that non-compliance with EPA requirements would provoke some form of disciplinary response. The experience of frontline workers, on the other hand, was that non-compliance was typically without consequences of any sort; it was, quite literally, inconsequential. While the EPA apparently had the capacity to ensure compliance with its rules, Texas City lacked either the resources or the will to ensure compliance with operating procedures.

More germane to the present discussion is the comparison between the attitude of managers to environmental regulations on the one hand and safety regulations on the other. Environmental law required rule compliance; safety law required risk management. Environmental law left little room for argument; safety law was inherently ambiguous and able to be contested. It was this contestability that enabled Texas City to avoid replacing the vent with a flare. This very contestability, then, was one of the causes of the Texas City accident.

---

18   Hale, op cit, p 57.

Risk management and senior executive commitment to safety

The failure to replace the vent with a flare highlights a crucial disconnect between the risk management approach and the safety commitment of the most senior company executives. Again, this is an important point that needs to be developed carefully.

Senior executives frequently say that safety is the top priority. For instance, one of BP's most senior executives, answerable only to the CEO of the entire BP group, wrote a two-page memo to all the business units in his far-flung empire about the need for cost cutting.[19] He specified among other things that training be deferred and recruitment frozen. He concluded the memo with this statement:

> "Clearly our highest priority is to ensure that we continue to go about our business safely."

In the context of the memo, the statement appeared like an afterthought that could only cast doubt on whether safety really was a top priority. But the problem with the statement runs much deeper. The problem is that it is not at all clear what it means to give priority to safety. Let me approach this issue by examining some of the statements made by this executive during his deposition.[20] He began as follows:

> "Never in my experience have we refused expenditure for safety purposes ... I'm not aware of any such conversation. I'm saying we didn't do it then and we don't do it now."

A little later, when commenting about budget cuts at Texas City, he said:

> "I cannot believe that they would have cut safety critical items, simply because ... it's not an acceptable thing to do ... they wouldn't have cut safety critical items, for sure. I'm sure of that."

When asked what he meant by "safety critical", he instanced certain kinds of training and the replacement of equipment that needs replacing — all things that had indeed been subject to cost cutting.

The statements by this executive are presumably sincere. But they are profoundly misguided. They evidence a dichotomous style of thinking that is at odds with reality. They assume that something is either safe or it is not, that certain items are either safety critical or they are not. The reality is that there is a continuum of risk and there is no point on this continuum that clearly separates the safe from the unsafe.

---

19   Manzoni, J, memo, 23 October 2002.

20   Manzoni, J, deposition, 8 September 2006, pp 7, 19, 34.

Nor does the risk management framework within which BP operated categorise things as safe or unsafe. The nearest it comes to this idea is acceptable versus unacceptable risk but, given that acceptability of risk depends to some extent on the cost of reducing it, this is no clearer than the distinction between safe and unsafe.

Let us apply this to the question of the flare. BP policy, like Amoco before it, was that flares were best practice, vents were not, and no new vents were to be built. But were vents "unsafe"? Certainly, flares reduce the risk and therefore flares are safer. Everyone could agree on that. But before the Texas City accident, at least, it was not easy to argue categorically that vents were unsafe. Not even OSHA had been able to maintain this claim when challenged.

What, then, is the relevance of the commitment of the BP senior executive in this context? If vents were unequivocally unsafe, his commitment would have meant something and he would have been committed to replacing them. But since safety decisions confronting managers are about how best to manage risks, not about whether things are safe or unsafe, there was no way that this commitment could be brought to bear. No matter how sincere this man's commitment to spending whatever it took to ensure safety, his commitment was essentially irrelevant when it came to questions such as whether to replace vents with flares.

There is one further aspect of the senior executive's position on safety that makes it essentially vacuous. Recall his statement:

> "Never in my experience have we refused expenditure for safety purposes ... I'm not aware of any such conversation. I'm saying we didn't do it then and we don't do it now."

The CEO of the entire BP group was similarly emphatic:

> "We are very clear: [none of my senior executives] ... or I have ever turned down a request for expenditure that has been identified as necessary for safety."[21]

These statements raise a crucial question: do these people ever see expenditure requests that are explicitly safety-related? Both men tacitly admitted that they don't. According to the former:

> " y expectation is that people successively up the system will budget for safety items first and then will do the discretionary items after that."

As a result, by the time spending requests reached his level, safety would not be explicitly identifiable, having been subsumed in broader expenditure headings. The CEO acknowledged this:

> "... the way I look at expenditure is on an aggregated basis, and I leave the details [to others] ..."

---

21   Browne, J, response to the Baker Report, 16 January 2007.

Clearly, then, claims by these men that they had never refused safety-related expenditure requests have very little meaning. To be quite specific, the question of whether to provide the capital needed to replace vents with flares was not the type of question that would ever have reached their level.

Budget priorities

A final argument I want to make here is that BP's budget priorities worked against risk reduction. This is a controversial claim so, again, it needs to be developed carefully.

There are numerous statements from BP plant managers that, when it came to budget priorities, safety was at the top.[22] After safety came reliability and sustainability of production, which obviously required spending on maintenance. Lowest on the list of priorities was spending to take advantage of emerging commercial opportunities.

But there were other statements of priorities that did not quite match this formulation. A capital investment manager said in his deposition that spending that affected the "licence to operate" was regarded as essential, and all other spending was discretionary.[23] The Baker Report speaks of BP's "three bucket" approach to budgeting, in which "licence-to-operate requests receive little push back; sustainability requests get reviewed by technical experts; and commercial requests are considered last and get the heaviest scrutiny".[24] An important implication of "license to operate" thinking is that it is not so much safety, as compliance with safety regulations, that is the top priority.

The argument can be taken a step further. It is not just compliance with safety regulations but compliance with government regulations in general that is necessary in order to maintain a licence to operate. Hence BP's striking commitment to comply with environmental rules. Indeed, the evidence is that, when Texas City discovered that it was not in compliance with benzene emission requirements, it was prepared to make exceptional budget requests to higher BP authorities, that is, requests for additional funds not foreshadowed in the annual budget. Moreover, those higher authorities were prepared to grant these requests in order to safeguard the licence to operate.[25]

It follows from this licence to operate philosophy that safety expenditure not specifically required in order to bring the site into compliance would not receive a top priority. Indeed, if an expenditure proposal was justified in terms of risk reduction rather than compliance with regulatory requirements, it is hard to see that this would be given any priority at all. Though continuous risk reduction

---

22   For example, MacLean, op cit, pp 11, 53.

23   Wundrow, W, deposition, vol 1, 5 June 2006, p 11.

24   Baker Report, p 83.

25   Hale, op cit, p 56.

was part of BP's stated philosophy, it had no place in BP's budget priorities, and risk-reduction proposals therefore stood little chance of success in the costcutting environment that BP had created at Texas City. Since any proposal to replace vents with flares was of this risk-reduction nature, BP's budget priorities would have rapidly eliminated it from consideration. It can reasonably be said, therefore, that BP's budget priorities served to undermine the risk management philosophy that it espoused.

This conclusion also helps to explain why BP failed to give priority to improving the instrumentation on the distillation column or even to repairing defective instrumentation. Such expenditure was not necessary to ensure the continuation of the licence to operate; it was not seen as necessary to ensure the reliability of production; and, finally, it was not necessary to take advantage of new commercial opportunities. As we saw in Chapter 2, better instrumentation would have prevented the Texas City accident. This is yet another way, then, in which BP's budget priorities contributed to the accident.

Conclusion

This chapter has sought to understand why it was that, on the day of the accident, combustible material escaped through a vent rather than a flare. Why was it that, despite widespread acknowledgment that flare technology was best practice, Texas City continued to use vents rather than flares? The answer is that the Texas City site operated according to a philosophy of risk management that did not specifically require vents to be replaced with flares. This might not have mattered if BP had implemented its policy of continuous risk reduction. It did not. Its focus was on doing whatever was necessary to safeguard its licence to operate. As we saw, this focus undermined the risk management approach.

oreover, the way BP's most senior executives categorised things as either safe or unsafe was fundamentally at odds with the risk management approach, making their commitment to safety essentially meaningless.

Chapter 4

Ignition

With a cloud of flammable gas at ground level, the scene was set for an explosion. All that was required was an ignition source. That source was provided by a vehicle parked 25 feet away with its engine idling. As the vapour cloud reached the vehicle, gas entered the air intake, causing the engine to speed up and backfire. Sparks shot out from the exhaust. Witnesses say it was this that triggered the explosion.[1]

Ignition was, of course, a critical step in the accident sequence. The fact that an ignition source was present is worthy of examination, yet the reports on the Texas City accident have very little to say about this issue. The 337-page report of the Chemical Safety and Hazard Investigation Board (CSB) devotes only two pages to it, and BP's own 176-page report spends less than a page on the topic and makes no mention of it in its list of "critical factors".[2,3]

This tendency to downplay the significance of the ignition source is itself a matter of interest. It is a tendency that can be observed in reports on petroleum industry accidents in general. For instance, in the Milford Haven report touched on in Chapter 2, the most detailed account of the ignition source is as follows:

"The released hydrocarbon formed a drifting cloud of vapour and droplets that found an ignition source 110 metres [away]."[4]

The thinking appears to be that there are so many potential ignition sources at a petrochemical site that, once a vapour cloud has formed, an explosion is almost inevitable. Efforts must therefore be focused on ensuring that flammable materials do not escape in the first place.

By contrast, inquiries into underground coalmine explosions go to a great deal of effort to identify the ignition source, weighing one theory against another to identify the most likely. They then frequently make recommendations aimed at eliminating, or least controlling, the ignition source that has been identified. The thinking in this case is that flammable gas is an ever-present problem and ignition control is therefore a vital safeguard against explosions.

---

1 *Investigation Report: Refinery Explosion and Fire* (CSB Report), Washington, US Chemical Safety and Hazard Investigation Board, March 2007, p 66; *Fatal Accident Investigation Report* (FAIR), London, BP, 9 December 2005, p 13.

2 CSB Report, pp 66, 67, 281, 282; FAIR, p 80.

3 FAIR, pp 80, 161ff.

4 Health and Safety Executive, *The explosion and fires at the Texaco Refinery, Milford Haven, 24th July 1994*, London, HSE, 1997, p 1.

However, to return to the petroleum industry, an explosion is not inevitable once a vapour cloud has formed. On at least six occasions in the 10 years prior to the Texas City accident, there had been accidental releases from the same processing unit that resulted in vapour clouds forming at or near ground level. As the CSB noted:

> "These ... releases could have been more serious if the vapour cloud had found a source of ignition."[5]

There is good reason, then, to try to control potential ignition sources at petrochemical sites, just as there is in underground coalmines.

This point was well understood by Esso at Longford. One of its ignition control strategies was to require employees, where possible, to move around the site on bicycles, rather than motor vehicles. This policy may well have stemmed from an Exxon accident years earlier in which an idling vehicle had ignited an explosion. I shall say more about this shortly.

Unfortunately, while Esso's policy was a sensible risk-reduction measure, it was not enough to prevent a vapour cloud finding an ignition source on the day of the Longford incident. Clearly, the highest priority in petrochemical industries must be to prevent any loss of containment in the first place. Ignition control policy is an additional risk-reduction strategy, not an alternative.

Texas City vehicle control policy

How, then, did Texas City deal with the fact that motor vehicles are a potential ignition source?

There was a formal site traffic safety policy in existence that stated:

> "For ... [maintenance shutdowns] and large capital projects, a traffic control plan must be developed ... **No vehicle must be left unattended with the motor operating. 'Unattended' is defined as no operator physically in the driver's seat of the vehicle in a position to control it.** *The provision does not apply to a vehicle providing a power source necessary for mechanical operation other than propulsion, passenger compartment heating or air conditioning; emergency response vehicles are considered power sources.*"[6]

Let us consider the meaning of this statement, noting that the italics and bold fonts have been added to assist the present discussion. A quick reading will no doubt focus on the sentences in bold and conclude that the vehicle that triggered the explosion had been left idling in contravention of this policy. But what is the significance of the sentence in italics? I have read it again and again and still cannot be sure of its meaning. I challenge readers to provide an interpretation. One possible interpretation is that where passenger compartments need heating

---

5   CSB Report, p 307.

6   FAIR, p 80.

or cooling, engines can be left idling, even when the vehicle is unattended. If so, the driver of the vehicle that triggered the explosion may well have been able to argue that he was not in violation of the policy. I do not wish to suggest that this is the correct interpretation — merely that, given the inherent obscurity of the statement, one could hardly blame a driver who interpreted it in that way. And what does it mean to say that emergency vehicles are a power source? Does it mean that they are exempt from the policy? Again, the sentence is so poorly written that it is impossible to say.

I have gone to some lengths to critique this policy because it provides an insight into the culture of casual compliance mentioned in Chapter 2. Where policies and procedures are written with such little consideration for those who must apply them, it is almost inevitable that they will be ignored or interpreted in ways that fail to take account of the hazards which the policies are intended to control.

The difficulties in the above statement suggest that no one had ever seriously tried to interpret the policy and no one had attempted to enforce it. If they had, the need to formulate the policy more clearly would have been painfully apparent. This interpretation is confirmed by managers who freely admitted at interview that there was no effective vehicle control policy at Texas City.

The first sentence in the quotation above has not yet been addressed. This places an obligation on management to develop a traffic control plan for maintenance shutdowns. Such a plan *had* been developed, but at least 55 vehicles were parked in the no access zone at the time of the accident, indicating that management had made no attempt to enforce its own policy.[7] It is clear that non-compliance at Texas City was not confined to frontline workers — managers were also implicated. This is an important observation, since it is further evidence, if any were needed, that the extent of non-compliance was far too widespread to be put down simply to the negligence or fault of certain individuals.

The universal failure to take the Texas City traffic control policy seriously is symptomatic of a lack of awareness of the risks associated with refinery processing. This profound blindness to major risk is something to which we shall return in later chapters.

Charlie Morecraft

One of the more disturbing features of this story is that it provides a particularly glaring example of the failure to learn lessons. Years earlier, Charlie Morecraft, a worker at an Exxon refinery in the United States, had left the engine of his vehicle idling, in violation of company policy, while he went to deal with a gas leak.[8] Something unexpected happened, and he was drenched in flammable material. He then realised that a vapour cloud was headed for his vehicle. He

---

7   CSB Report, p 141. The figure 55 is mentioned in a CSB release dated 27 October 2005.

8   Website at www.safetyworld.com/FLBook/fl_book_4_9.htm.

sprinted away but saw his truck explode in a ball of flame that engulfed him. The refinery went up in flames and Charlie was burnt almost to death. He wanted to die, but he survived and underwent years of excruciatingly painful rehabilitation.

Since his recovery, Charlie has commanded huge fees speaking to audiences in the US and around the world about his ordeal and about his stupidity in failing to comply with safety procedures, in particular, the requirement not to leave a vehicle engine idling. He has spoken to many petroleum industry groups, including BP audiences, and he lists BP among his clients on his website. There is no doubt that his message is well known within BP, but it had no apparent impact on behaviour at Texas City.

Chapter 2 identified a number of examples of BP's failure to learn lessons from elsewhere in the industry. The tolerance of idling engines at Texas City is perhaps the most striking example of this failure. Despite the ongoing publicity that surrounds Charlie   orecraft and his story, Texas City remained unaffected. This failure to learn has been so consistent and so systematic that Texas City can reasonably be said to have suffered from a learning disability, as I shall argue in more detail later.

Conclusion

Given that the main reports on the Texas City accident contain very little analysis of why an ignition source was present, this chapter has been necessarily brief. Ignition is, however, an essential element in the chain of causation: had no ignition source been present, the accident would not have happened. The matter therefore deserves attention in its own right.

Steps taken at Texas City since the explosion vindicate this conclusion about the need to focus on the risk of ignition. BP has introduced a site-wide transportation system that has removed more than 500 vehicles from the site.[9]

The present chapter has served to deepen several of the themes that run through this book. First, it provides a mini case study of the problem of non-compliance and the reasons for it. This was not just the behaviour of frontline workers — managers were also implicated. Second, it provides a further example of the pervasive lack of risk awareness at Texas City. Third, it highlights the learning disability from which Texas City suffered. These are all matters that require further explanation.

---

9   Broadribb, M, *Three years on from Texas City*, in the proceedings of the US Center for Chemical Process Safety 23rd International Conference, New Orleans, April 2008, p 6.

Chapter 5
Why so many deaths?

The ignition of the vapour cloud resulted in a massive explosion and fire that did great damage to everything in the vicinity. But explosions do not necessarily result in death. The Milford Haven explosion (initiated by the overfilling of a distillation column, exactly as in the Texas City case) killed no one, despite destroying large sections of the plant. In contrast, the Texas City explosion killed 15 and injured nearly 200 people. This was an exceptional toll, due to the presence of a large number of people who were not essential to the work being undertaken. The question that this chapter addresses, then, is: why were so many non-essential people located so close to hazardous equipment?

All of the dead and injured were engaged in the maintenance work being carried out on a nearby but unrelated processing unit. They were housed in a series of so-called trailers (mobile or portable office blocks). The trailer closest to the vent, a double-width trailer, was a mere 120 feet away. It normally housed 13 people, but at the time of the accident there were 22 people gathered inside for a weekly meeting. Twelve of the 15 dead were killed in or around this trailer. The other three deaths occurred in a trailer that was 136 feet from the vent. Both of these trailers were demolished by the explosion.[1]

The trailers did not need to be where they were. They could have been located hundreds of feet further away — in which case, no one would have died. They were where they were principally for reasons of convenience. For nearly 30 years, trailers had been located in this area during maintenance shutdowns, as close as possible to the site of work. And for many years, no one had questioned the wisdom of locating so many people in temporary accommodation so close to hazardous equipment.

All this began to change following the promulgation of the United States Government's process safety management standard in 1992, which required, among other things, that a risk assessment be carried out for facility siting, including the siting of temporary trailers. In 1995, Amoco published guidelines for siting various types of buildings and, in 1999, Texas City adopted what it called a "management of change" process that governed the siting of trailers.

A management of change process was carried out for the double-width trailer prior to the commencement of the maintenance work. We now know that it was entirely inadequate. To understand why, we must examine the process in some detail.

---

1 Details from the *Investigation Report: Refinery Explosion and Fire* (CSB Report), Washington, US Chemical Safety and Hazard Investigation Board, March 2007, p 257, plus an analysis of trailer damage on the CSB website.

The groundwork had been laid by Amoco in 1995. Following detailed calculations, risk engineers had concluded that wooden trailers, such as the one in which so many people died, needed to be at least 350 feet from potential explosion sources in order to be safe. This was, in effect, a general rule that Amoco had devised for itself. But it was not an inflexible rule. Trailers could be located closer than 350 feet, provided a risk assessment was conducted that took account of the specific site. I shall examine each of these steps below. Together they demonstrate the risk management approach in action, and together they demonstrate various limitations of this approach. This chapter, then, continues a theme introduced in Chapter 3.

The 350 feet rule

The risk assessment that generated the 350 feet rule depended on various assumptions that turned out to be inapplicable in the Texas City case. I mention two in what follows.

First, vapour cloud explosions often occur in spaces that are congested with pipes, tanks and other equipment, and the intensity of the shock waves that they generate is greatly affected by the amount of congested space; the larger the amount of congested space, the greater the shock wave. In their calculations, the risk engineers used the average size of congested spaces across several North American refineries. However, the congested spaces at Texas City are larger than this average. The result is that the risk calculations underestimated the intensity of shock waves that would be produced by explosions at Texas City.[2]

Second, the risk engineers made assumptions about the vulnerability of various structures to shock waves of a given strength. Perhaps surprisingly, they assumed that wooden trailers were not as likely to collapse as steel, concrete or brick structures. So, while concrete and brick buildings had to be at least 700 feet from potential explosion sources and steel-framed buildings at least 450 feet, trailers could be located as close as 350 feet. The rationale was as follows:

> "Data from actual events indicate that trailers tend to roll in response to a vapor cloud explosion and walls and roofs do not collapse on occupants, resulting in fewer serious injuries/fatalities."[3]

This is a perplexing rationale. The general duty in the federal *Occupational Safety and Health Act* is a duty to prevent "death or serious harm" and, from this point of view, it is reasonable to focus on the probability of death or serious injury. However, anyone inside a trailer that rolls over is quite likely to be injured, if only slightly. To ignore this possibility hardly seems sensible. At any rate, the point needs to be made that the rule about trailer siting took no account of the possibility of less serious injuries.

---

2   *Fatal Accident Investigation Report* (FAIR), London, BP, 9 December 2005, p 102.

3   FAIR, p 102.

Quite apart from this, there were particular circumstances at Texas City that invalidated the assumption about the vulnerability of trailers relative to other structures. Texas City is in a hurricane-prone zone and trailers are normally tied down as a precaution. They will therefore "flex less or roll less in response to explosions".[4] In the case of Texas City, therefore, there was no reason to allow trailers to be closer to explosion sources than fixed buildings.

The strongest argument against the various assumptions made about the vulnerability of trailers is retrospective: it comes from a careful analysis by the Chemical Safety and Hazard Investigation Board (CSB) of the actual damage that they suffered. The CSB found that trailers had been totally destroyed by blast pressures that were predicted to do only very minor damage.[5] In this matter, then, the assumptions made by the risk engineers turned out to be grossly wrong.

Another assumption made by the risk engineers concerned acceptable levels of injury. They decided that the trailers would be regarded as being at a safe distance, provided less than 10% of the occupants were likely to be killed or seriously injured in the event of an explosion.[6] Various ethical and practical objections can be raised to the idea of acceptable risk, but these will not be canvassed here.[7] The main point to note is that, on close examination, the rule did not presume that people located at 350 feet or beyond would be absolutely safe, merely that the risk at that distance was acceptably low.

There is an important conclusion to be drawn from this discussion. The validity of a risk assessment depends crucially on the validity of the assumptions that are made by the risk engineers. If those assumptions turn out to be wrong, or at least not applicable to the particular case at hand, then the results of the risk assessment (in this case, a minimum safe distance) will be invalid. What this means is that no risk assessment should be accepted at face value without a consideration of the assumptions that went into it. In fact, critical personnel at Texas City had no idea about the assumptions that went into this calculation. The person who led the specific trailer siting risk assessment process was asked at interview to explain the 350 feet figure. All he could say is that "350 feet . . . is the number that someone came up with".[8] A close examination of the assumptions made by the risk engineers could conceivably reveal that they result in an overestimate of the true risk, but certainly at Texas City, they resulted in an underestimate. As the BP report itself concluded: ". . . the minimum safe distance

4   Ibid.

5   CSB Report, p 129. The CSB Report (p 128) identifies several other assumptions that resulted in an underestimate of the impact of blast pressure on trailers.

6   FAIR, p 102.

7   For an account of these problems, see Hopkins, A, *Safety, culture and risk*, Sydney, CCH Australia Limited, 2005, Ch 12.

8   Steele, W, deposition, 2 February 2006, p 55.

for trailers at Texas City Refinery should be greater than the recommended 350 feet."[9]

However, despite the inadequacies of the 350 feet rule, it must be noted that, had the rule been followed, there would probably have been no deaths and the injury toll would not have been as severe. There were various trailers located at 350 feet and beyond, and the level of damage that they suffered fell far short of the total destruction experienced by the double-width trailer.[10]

The Texas City risk assessment

Trailer siting at Texas City was not determined by a simple application of the 350 feet rule — although this came into it, as we shall see in a moment. What was required was a "management of change" process, the change in this case being the movement of trailers onto the site. It was carried out by a group of individuals who were scheduled to take part in the forthcoming maintenance work. Their preliminary task was a "brainstorming activity" to identify any matters of concern that group members might have about the proposed trailer siting.[11] They came up with two. The first was the proximity of an oily water separator which might, in certain circumstances, produce very small quantities of flammable vapour.[12] The proposed risk reduction was to check electrical sources in the area and to signpost evacuation routes in case a rapid evacuation should be necessary. A second risk was the possibility of a forklift vehicle hitting a pedestrian. This was identified because of a recent forklift fatality at another BP site.[13] But nothing was done about the signposting of evacuation routes or the forklift traffic risk. Although they were listed as "action items", they were never in fact actioned.

The Texas City policy specified that the management of change process should be commissioned by an operations superintendent who was required to ensure that all action items arising were finalised before giving his approval for the trailers to be occupied. However, not only were the above two action items never finalised, but the superintendent never gave his approval for the trailers to be occupied.

---

9   FAIR, p 102.

10   CSB Report, p 270. See also the study of trailer damage on the CSB website. One metal trailer was located at 349 feet and both its occupants were knocked unconscious by the blast and suffered broken bones. Workers were injured in trailers as far away as 480 feet (CSB Report, p 241).

11   Steele, op cit, p 52.

12   Steele, op cit, p 53; FAIR, p 94.

13   Steele, op cit, p 53.

These procedural failures provide further evidence of the attitude of casual compliance discussed earlier and, importantly, of the way in which management was implicated. It appears that quite senior managers were aware of the procedural requirements but failed to implement them.[14] The question of why they failed in this way must be deferred till later.

Lawyers for those suing BP pounced on this failure to follow through on the formal procedures, seeing it as contributing to the massive death toll. But it has to be said that, had the remaining action items been completed and the occupancy of the trailers authorised in accordance with Texas City policy, the toll of death and injury on the day of the accident would have been no different. It was the location of the trailers that was the problem and this had not been questioned by any of the participants in the management of change process. Somehow, the whole management of change process had simply missed the point.

Bypassing the 350 feet rule

Just how the point was missed becomes obvious when we examine the next step in the management of change process, which required the decision-making group to answer a complex series of questions set out in a 40-page *Facility Siting Screening Workbook*.[15] One of the early questions in the workbook asked whether the proposed location was less than 350 feet from a process unit. The team answered "yes". Actually, there were *two* units within 350 feet, and the process unit identified by the team was *not* the one involved in the accident. The vent that emitted the vapour cloud was only 120 feet away, but this fact was not recognised by anyone at the time. I shall return to the significance of this mistake shortly.

The management of change process at Texas City did not require the 350 feet rule to be observed. But it did require that, if trailers were to be located closer than 350 feet to a process unit, the team proceed further through the workbook in order to evaluate the explosion risks. In principle, this necessitated another set of risk calculations, similar to those from which the 350 feet rule was derived. However, the people engaged in the risk assessment for the Texas City trailer siting were not risk engineers and had not been trained in the use of the workbook.[16] They were therefore unable to complete the required analysis. Apparently, though, they did not regard this as a significant problem. Their implicit assumption was that the double-width trailer and all other trailers were to be located where they had always been located, and that the purpose of the exercise was to ensure that this was done as safely as possible. They were satisfied that they had done this.

---

14   FAIR, pp 92, 93.

15   Sorrels, S and Lash, W, *Facility siting screening workbook*, Amoco, April 1995.

16   CSB Report, p 126.

The failure of this group to take the possibility of a vapour cloud explosion seriously is readily apparent in the documentation. At one point, the workbook asked whether the group had evaluated "personal work areas within the building ... located along walls facing potential blast sources".[17] This was a crucial question; it specifically directed their minds to the risks of an explosion emanating from a nearby process unit. If the answer was "no", that is, they had not evaluated these risks, then the trailers would have to be located elsewhere.[18] So it was that the group unanimously agreed to answer "yes" when, on any reasonable interpretation, the answer should have been "no".[19]

In summary, the group bypassed the 350 feet rule but failed to perform the risk assessment that should have been carried out in these circumstances.

The two levels of risk assessment reconsidered

The preceding discussion identifies two levels of risk assessment: one at the company level and the other at site level. The comparison is instructive. It turns out that the site risk assessment was inherently biased in a way that the company assessment was not. The aim of the company-level risk assessment was to devise a rule about where trailers could be safely located. Risk varies continuously with distance: the further away the trailers, the lower the risk. The risk assessment needed to convert this risk continuum into a decision-making dichotomy: trailers may be located this close (350 feet) and no closer. As we saw earlier, there is a degree of arbitrariness involved in the determination of this rule, but the process is not essentially biased in favour of any particular outcome, that is, any particular distance. On the other hand, a site-level risk assessment of the type described above starts with a proposed location and then seeks to determine whether this location is justified. The process inevitably puts pressure on risk assessors to find in favour of a proposed location. This is the so-called confirmation bias.

A second point of contrast is that, while the company-level risk assessment was done by risk engineers, the site-level risk assessment was done by people who were incapable of carrying it out effectively. It is almost inevitable that local risk assessments of this nature will be carried out by people who are not competent in quantitative risk assessment. This can only reinforce their tendency to fall prey to a confirmation bias.

None of this is to suggest that the 350 feet rule was adequate; it clearly wasn't. But, just as clearly, it was a great deal better than no rule at all. We come, then, to the following conclusion. The best way to ensure safety in matters such as trailer location is to devise a rule beforehand, rather than allowing decisions to be made on the basis of individual risk assessments. The latter approach is likely to introduce confirmation bias and to result in less competent risk assessments. Had

17   FAIR, p 95; Steele, op cit, pp 62, 63.
18   Steele, op cit, p 63.
19   Ibid.

Texas City stuck to the 350 feet rule rather than allowing case-by-case risk assessments, the outcome would have been very different.

This conclusion echoes that of Chapter 3 in which it was argued that, had there been a rule-based approach rather than a risk management approach to the question of replacing the vent with a flare, the tragedy at Texas City would most likely have been averted.

A further example of converting risk management into rule compliance

The desirability of converting risk management into rule compliance has been recognised in other contexts. For instance, a recent study at a nuclear power station examined how managers make decisions that require balancing production against safety.[20] A particular issue they faced was what to do when some backup safety system failed; for example, when an emergency backup pump was found to be non-operational. It might be one of several redundant backup systems, but to continue production in these circumstances meant operating with a smaller safety margin. Should the process be closed down or was it reasonable to accept the slightly higher risk and continue operating while the issue was being resolved? The problem was that it was seldom possible to know how long this would take. Managers knew that, at any one point in time, a few more hours operating with the smaller margin of safety would not increase the risk appreciably. Accordingly, if the question was whether to stop now or to continue operation a little longer in the hope that the problem would soon be fixed, it was reasonable to continue.

However, managers knew that the longer they continued operating in a degraded state, the greater the tendency to "normalise" the situation, that is, to accept the greater level of risk as normal. The normalisation of risk has been a significant factor in many major accidents. For example, prior to both the Challenger and Columbia space shuttle accidents, a certain level of equipment malfunction came to be accepted as normal because it had not in the past led to disaster. People became desensitised to the risks of operating in this way. Ultimately, these malfunctions proved fatal.[21]

The way the nuclear power station managers dealt with their dilemma was to draw what they called "a line in the sand": if the matter was not resolved within, say, 24 hours, they would stop production. In this way, they created a rule for themselves: if 24 hours passed without a resolution, there was no longer a risk assessment to be carried out, there was a rule to be complied with, and the decision was clear-cut.

---

20   Hayes, J, Canberra, Australian National University (PhD thesis in progress).

21   Columbia Accident Investigation Board, *The Report*, vol 1, Washington, National Aeronautics and Space Administration, August 2003; Vaughan, D, *The Challenger launch decision: risky technology, culture and deviance at NASA*, Chicago, University of Chicago Press, 1996.

One is reminded of the story of the frog in a pot of water that is being slowly heated. At any particular point in time, a few seconds more will not make an appreciable difference to the temperature of the water and hence to the risk of death.   oreover, as the temperature increases, the frog progressively normalises it. This is an approach that eventually proves fatal. Such a frog would be well advised to draw a metaphorical line in the sand and say to itself, "when the temperature reaches 50 degrees, I'm out of here".[22]

There are, no doubt, many circumstances in which it is appropriate to convert risk management into simple rule compliance. It is not within the scope of this book to pursue this question, but it is clearly something that deserves serious consideration, both by regulators and by companies themselves.[23]

The failure to recognise the vent as a potential explosion source

One matter has been left hanging in the preceding account of the trailer siting risk assessment: the failure of the team to identify the vent involved in the accident as a potential explosion source. Given this failure, even if team members had been committed to the 350 feet rule, they might, theoretically, have placed trailers within 350 feet of the vent without realising that they were violating the rule. In other words, even a commitment at the site to the 350 feet rule may not have prevented the tragedy that occurred. However, the particular layout of the Texas City site meant that any suitable trailer location which was at least 350 feet from the identified process unit would almost certainly have been at least 350 from the vent as well. The team's failure to identify the vent as a potential explosion source would thus not have mattered.

The failure of the team to recognise the potential danger of the vent deserves some further explanation. The dangers associated with *flares* were well understood at Texas City. Various people gave evidence that they had witnessed fire cascading down from flare towers.[24] This was why flares were located in flare

---

22 Another example of this line-in-the-sand strategy is the rule of three that Shell has adopted for on-the-spot decision-making (Hudson, P, et al, *The rule of three: situation awareness in hazardous situations*, in the proceedings of the conference of the Society of Petroleum Engineers, Caracas, June 1998; see also Shell's *Hearts and minds programme*, available at www.energyinst.org.uk/ heartsandminds). The principle is that, if three risk-enhancing factors are present, the activity should be suspended or avoided. Putting this metaphorically, three simultaneous amber lights should be treated as the equivalent of a red light. For example, if I have had one drink (but I am not over the legal limit), I am tired, and visibility is limited because it is a wet night, then I should not drive myself home.

23 The clearest example of turning the risk continuum into a decision-making dichotomy is the idea of specifying a number that marks the boundary between acceptable and unacceptable levels of risk. However, the major practical problem with this strategy is that it is virtually impossible for on-the-spot decision-makers to calculate the numerical level of risk they confront, so as to be able to compare it with the risk-acceptability standard. Quantitative risk assessment is therefore of little or no use for on-the-spot decision-makers.

24 Maslin, P, deposition, 13 July 2006, pp 27, 30; Hoffman, M, deposition, 2 August 2006, p 70.

yards. No one would have authorised the placement of a trailer in a flare yard, and one manager gave evidence that he had once discovered a trailer in a flare yard and ordered its removal.[25] But none of the people who had seen fire cascading from a flare tower had witnessed a vapour cloud explosion resulting from an accidental release from an unflared vent.

On at least six occasions in the preceding 10 years, there had been an unplanned release from the vent in question that resulted in vapour clouds forming at or near ground level. All of these releases were potentially very dangerous occurrences. Generally speaking, though, while a vapour cloud can be more deadly, it is not as dramatic as a cascading fire. As a result, people were not as aware of the potential dangers.[26] So it was that, while flares were isolated in fenced-off yards, the potentially more deadly vents were not.

None of the members of the trailer siting management of change team had any involvement with the process unit where the accident occurred, and none knew of the previous vapour cloud incidents.[27] The team leader said at interview that, had he been aware that the vent sitting just 120 feet from the proposed trailer location had experienced so many dangerous near misses, he would most probably not have agreed to locate the trailer at that site.[28] This failure to take account of previous incidents in a risk assessment exercise is something to which I shall return in Chapter 6.

A warning ignored

At least one individual at Texas City was concerned about how close the trailers were to the process unit that was involved in the accident. This man was an instrument and electrical technician who had worked on the unit in question and in particular on the instrumentation on the distillation column. He was concerned about the condition of this unit, having previously discovered three major leaks.[29] He described it as a "piece of junk". This man was aware of the approaching startup of the unit and was worried. As he said at interview:

> "A startup is the most critical time of a unit because you have got flows, pressures and temperatures that are swinging one way or another and the unit's unstable. And [this unit] . . . looked pretty ratty, looked like it needed a lot of work . . . When a unit looks like that, and they start it up, you are pretty cautious. Any unit you are cautious but . . . a little alarm goes off in your head saying you want to be a little more cautious with a unit like this when you are starting it up."[30]

---

25   Hale, R, deposition, 1 June 2006, p 75.

26   Ibid, p 76.

27   FAIR, p 93.

28   Steele, op cit, p 82.

29   Runfola, J, deposition, 22 May 2006, p 15.

30   Ibid, p 22.

Like others at the site, he was aware of the dangers of flares but unaware of the specific risks posed by vents. His concern was simply that trailers were located too close to a unit that was about to be started up. He had worked at other sites where trailers had been kept well way from process units and he believed that this should be the practice at Texas City as well. Accordingly, he expressed his concerns to one of the senior managers, a man who reported directly to the Texas City manager. The senior manager made some inquires and got back to the technician saying that the trailer siting had been subject to the management of change process and that it was therefore out of his hands.

To his credit, the technician did not accept this reassurance. A few days prior to the startup, he happened to be in one of the trailers. He told the occupants that the process unit in question was "pretty raggedy". He went on:

> "It would be a good idea if y'all contacted operations to find out when they are going to start this unit up because you don't want to be here during the startup."[31]

Here, then, was a very specific warning. This man understood the situation with remarkable clarity. He recognised that startup was a period of heightened risk and he was aware that this startup was particularly dangerous because of the poor condition of the process unit. But his was a lone voice. He was surrounded by people and practices that were not attuned to explosion risks. He was up against an organisation that displayed very little awareness of the most significant hazards on site.

Consequence-based decision-making

The obvious problems with risk-based decision-making in relation to the trailer siting at Texas City have led BP to adopt a new approach.[32] It is described as a shift from risk-based to consequence-based decision-making. On closer examination, this new approach turns out to be a shift in the direction of the rule-based strategy advocated above.

There are two relevant features of the risk-based approach. First, it is one that holds that some level of risk is inevitable and that the real question is: how much risk is acceptable? Second, the risk-based approach treats risk as a product of likelihood and consequence, that is, the more likely the event, the greater the risk, and the more severe the possible consequence, the greater the risk. One result of this way of thinking is that, even though the consequences of an event may be severe, if the likelihood is assessed as extremely low, then the risk is low. Furthermore, if the risk is low enough, it may be deemed acceptable.[33] In the present context, a risk-based approach legitimates placing people in buildings that are not designed to resist blast shock waves, provided the likelihood of an

---

31 Ibid, p 23.

32 Maslin, op cit, pp 56, 57; Sorrels, S, deposition, 8 June 2006, pp 28, 29.

33 Amoco used "annualised" occupancy data to decide whether facility siting risks were acceptable. This was problematic, as discussed in Appendix 1.

explosion is low and the expected number of deaths, in the unlikely event of an explosion, is low.[34]

The consequence-based approach, on the other hand, takes no account of likelihood and repudiates any idea of acceptable risk. The philosophy is that, if the consequences are severe, then people must be protected from them, no matter how unlikely they may be. Decision-making must ensure that people are at all times protected from death or serious injury. The principle, here, is that any building that is to be located in a potential blast zone must be built to withstand the blast pressure. To implement this principle, BP decided to divide each refinery into zones, based in part on calculations of the possible blast pressures: a red zone, in which no temporary buildings may be located; an orange zone in which temporary buildings may be located if they are designed to withstand the blast pressures that are possible in that zone; and a yellow zone beyond that.[35] Clearly, this policy is about eliminating risk, not reducing it to an acceptable level. Note that it creates a rule that must be followed, rather than leaving it up to local managers to make individual risk-based decisions about trailer siting.

The principle of consequence-based decision-making is not one that can be universally applied. In many, if not most, situations, it is not possible to protect people from all risk. In the case of trailer siting, however, the risk to trailer occupants *can* be eliminated by remote siting. What makes it possible to adopt the principle in this case is that the financial cost of moving trailers out of range is not significant; all that is really at stake is questions of convenience.

So far as reasonably practicable

BP's move to consequence-based decision-making for trailer siting can be conceptualised in a different way, namely, as a move to reduce the risk "so far as reasonably practicable". This is a crucial idea, since it is precisely what is required by occupational health and safety law in the United Kingdom, Australia and certain other countries.[36] It is not required by the federal *Occupational Safety and Health Act* in the US. This is not the place for an extended discussion of the meaning of reasonable practicability, except to note that one relevant issue is cost. If the cost of a possible risk-reduction measure is insignificant, then it follows that it is reasonably practicable to implement this measure, even if the risk is already low. From this point of view, there is no way of determining an acceptable risk figure beforehand; it will always depend on how costly it is to further reduce the risk. For instance, the level of risk associated with space shuttle flight is way above the level deemed acceptable in most other situations. It may nevertheless be as low as reasonably practicable because the cost of making shuttle flight safer would be prohibitive. On the other hand, no matter

---

34 This philosophy is clearly articulated in a BP training document (BP Group HSE standard, *Major accident risk-awareness training*, 17 October 2002).

35 Maslin, op cit, p 56. See also *The Report of the BP US Refineries Independent Safety Review Panel*, Washington, US Chemical Safety and Hazard Investigation Board, January 2007, p 53.

36 Robens-inspired jurisdictions.

how low the level of risk involved in siting trailers at, say, 350 feet from process units, this is not as low as reasonably practicable because the trailers can be sited further away without significant expense. In short, the trailer siting case highlights an inconsistency between the idea of acceptable risk on the one hand and the idea of reasonable practicability on the other. The risk at a particular trailer location may be below some previously agreed acceptable level, but it may still not be as low as reasonably practicable. What this means is that reducing risk to some predefined acceptable risk level does not automatically assure compliance with the laws in the UK and Australia, among other places. This is a major challenge to the whole philosophy of acceptable risk, but it is one that to date has been barely recognised.[37]

In summary, BP's move to consequence-based decision-making in the case of trailer siting can be seen as a move to reduce risk as far as reasonably practicable. It is a philosophical shift that brings it into alignment with the law in the UK, where BP is based.

Conclusion

The risk assessment exercise for the trailer siting at Texas City was a debacle. The people concerned did not understand the major hazards that they faced and they did not recognise that the essence of the exercise was to evaluate the explosion risks associated with the particular location. In any case, they lacked the necessary technical competencies for this task. The procedure enabled them to bypass the 350 feet rule devised by Amoco's risk engineers and to substitute their own judgment without applying any of the rigour that went into formulating the original rule. The outcome demonstrates how risk assessments about major hazards can go totally awry when they are made at the local level, under the influence of local pressures, and without scrutiny by more competent people. In the case of major hazards at least, the risk assessment process must be transformed into a rule-compliance requirement by the time local decision-making comes into play.

This is not to say that the 350 feet rule was adequate. Compliance with this rule would not have prevented serious injuries, although it would probably have prevented all of the fatalities. It is important to realise that the validity of such rules depends critically on the assumptions that are made for the purposes of calculations, and to be aware that these may be in error, or at least inapplicable, in particular circumstances. In short, there is a need to maintain a level of scepticism even about risk assessments made by competent people.

BP appears to have drawn a further conclusion: where reasonably practicable, the philosophy should be one of eliminating risk, rather than reducing it to any predetermined acceptable level. This is precisely what the law in some countries requires.

---

37   This issue is dealt with more fully in Hopkins, A, *Safety, culture and risk*, Sydney, CCH Australia Limited, 2005, pp 120–122.

Chapter 6
Blindness to major risk

Previous chapters followed the accident sequence quite closely and explored some of the specific reasons for the explosion and massive loss of life. At several points, I identified blindness to major risk as a root cause of sorts. I say "of sorts" because, strictly speaking, there are no root causes, only points at which we stop asking "why". Too often, we stop prematurely. The purpose of this chapter is to ask why it was that Texas City seemed to be relatively unconcerned about the major hazards on site that had the potential to blow the place apart. How are we to understand this blindness to major risk?

Process safety and personal safety

The answer to this question has to do with the way that safety was understood at Texas City as personal rather than process safety. This distinction was highlighted in all of the reports on the Texas City accident, but was given greatest prominence in the Baker Report, which expressed it as follows:

> "*Personal or occupational* safety hazards give rise to incidents — such as slips, falls, and vehicle accidents — that primarily affect one individual worker for each occurrence. *Process* safety hazards give rise to major accidents involving the release of potentially dangerous materials, the release of energy (such as fires and explosions), or both. Process safety incidents can have catastrophic effects and can result in multiple injuries and fatalities, as well as substantial economic, property, and environmental damage. Process safety in a refinery involves the prevention of leaks, spills, equipment malfunctions, over-pressures, excessive temperatures, corrosion, metal fatigue, and other similar conditions."[1]

The term "process safety" originates in the United States; in some other parts of the world, process safety is referred to as asset integrity or technical integrity. Furthermore, in some contexts, process safety hazards are referred to as major hazards or major accident hazards.[2] These latter terms are more general in that they also apply to industries such as rail and air transport and underground mining, all of which can experience catastrophic accidents. The only drawback with the term "major hazard" is that it suggests, misleadingly, that other types of hazard are minor. The hazards of high-voltage electricity, or working at great heights, are potentially fatal and cannot be described as minor. Indeed, many

---

1 *The Report of the BP US Refineries Independent Safety Review Panel* (Baker Report), Washington, US Chemical Safety and Hazard Investigation Board, January 2007, p x.

2 For example, in jurisdictions that require safety cases.

organisations describe near misses associated with such hazards as "high potential incidents" or HiPos. In short, use of the term "major hazard" to describe certain types of hazards must not be understood as implying that other types of hazard are insignificant.

The above quotation provokes one final observation on terminology. There is an assumption in the first line that "personal safety" and "occupational safety" are the same. This is problematic. All hazards at work affect occupational safety, whether they be personal or process safety hazards. For this reason "personal safety" is to be preferred to "occupational safety".

Since personal safety and process safety are concerned with different kinds of hazards, it is logically possible to focus on one type of hazard and not the other. This was the essence of the problem at Texas City: the focus was on personal safety hazards, not process safety hazards. This was not necessarily a conscious choice. Many managers did not understand the distinction or, if they did, they assumed that attention to personal safety hazards would automatically ensure that attention was given to process safety. So, the Texas City manager said at interview:

> "I connect them in my mind . . . In my mind you don't have good [personal injury statistics] and poor process safety or vice versa . . . they're integrated. I don't differentiate them in my mind."[3]

To his credit, this man was very concerned about personal safety at Texas City and was doing a great deal at the time of the accident to improve personal safety. "Changing the safety culture at the plant was my number one priority", he said.[4] But his failure to distinguish between personal safety and process safety meant that process safety did not get the attention it deserved.

His attention to personal safety was reinforced by the concerns of his superiors. He described at interview how he had been summoned to London, five months before the explosion, to talk about a fatality that had just occurred at Texas City. He said he thought his career was over at that point. The discussion was all about personal safety; process safety was not mentioned, he said.[5]

The focus on personal safety was driven by the way safety was measured in terms of total "recordable" injuries. A recordable injury is defined by the US Occupational Safety and Health Administration (OSHA) as an injury that results in death, days away from work, alternative work duties, or medical treatment beyond first aid.[6]

---

3   Parus, D, deposition, vol 2, 10 July 2006, p 45; see also Ralph, W, deposition, vol 1, 2 March 2006, pp 8, 43.

4   Parus deposition, op cit, p 23.

5   Parus, D, interview for Bonse accountability project, 12 October 2006, p 10.

6   OSHA Standard 29 CFR 1904.7: *General recording criteria*. This is a slightly simplified version of the OSHA definition. Illnesses that meet the same criteria are also recordable.

Texas City was doing very well by this measure; its OSHA recordable injury rate was at an all-time low.[7] Moreover, the rate was one-third the industry average.[8] In the months before the accident, senior BP executives congratulated the Texas City workforce on its achievements in this respect and increased bonus pay in recognition.[9] There had been three fatalities at Texas City in 2004, two of them arguably related to process safety issues, but these fatalities apparently caused no one to question the significance of the injury rate achievement. Injury rate was king; it meant more than any other indicator. It was this figure that was publicised, it was this measure that was used for making comparisons both within the company and with other companies and, finally, it was performance with respect to this measure that was rewarded financially.

The problem with using injury statistics as a measure of safety is that most injuries are caused by personal safety hazards. Years may go by without an injury that is attributable to a process safety failure, and a plant may go for many years without the sort of process safety incident that gives rise to multiple fatalities. So, if an organisation is seeking to drive down its injury rate, it will naturally focus on the hazards that are contributing to that rate on an annual basis. These may be vehicle hazards, working at heights, failure to wear personal protective equipment (PPE), and so on. The stronger this focus, the more likely it is that the organisation will become systematically complacent with respect to major hazards, precisely because they do not contribute to the recordable injury rate on an annual basis. This is what happened at Longford, and it has been identified as an issue in many recent major accident investigations.

A vignette

There is one episode at Texas City that illustrates very poignantly the discrepancy between the two types of safety. Just over half an hour before the explosion, a meeting of about 20 people was held in the control room. It broke up 10 minutes before the explosion.[10] None of the meeting participants was aware of the problems that were unfolding as the meeting progressed. It was suggested at one stage that this meeting might have distracted the control room operator at a critical time. However, the Chemical Safety and Hazard Investigation Board (CSB) concluded that this was not the case, since it was decision-making prior to the meeting that led to the distillation column overfill.

---

7   Parus deposition, op cit, p 16.

8   *Investigation Report: Refinery Explosion and Fire* (CSB Report), Washington, US Chemical Safety and Hazard Investigation Board, March 2007, pp 202, 203.

9   CSB Report, p 176. See also Parus, D, memo, 17 March 2005.

10   CSB Report, p 304.

But there is more to be said on this topic. Such meetings are often held in refinery control rooms, so that control room operators can be present even while they are on the job. This is a revealing practice. It is evidence of a belief that, whatever the operators are doing, it is not so critical as to require their full attention. Even at startup, the time of greatest danger, it seems that no one at Texas City gave any thought to the possibility that the control room operator might need to give his full attention to his job and that the meeting might constitute a distraction.

This is in striking contrast to what goes on in air traffic control rooms, where the environment is hushed and all unnecessary conversation is discouraged. The assumption is that air traffic controllers must be able to give their undivided attention to what they are doing. No one would dream of convening a meeting in this setting. It is true that air traffic control is a more critical activity, in the sense that air traffic controllers are far more active players in what they do than refinery control room operators are. Furthermore, situations can alter and get out of hand much more rapidly. Nevertheless, the contrast highlights the relatively casual attitude to major accident risk that existed in the Texas City control room.

But the real point of this story lies in the purpose of that control room meeting. It was to celebrate safety. A 35-day maintenance shutdown for two other process units on site had just been completed without a single recordable injury and with only two first aid treatments.[11] Achievements like this are indeed worthy of celebration, and such celebrations are an excellent way to reinforce the importance of safety. But the celebration was about personal safety only, and participants were oblivious to process safety failures that were about to lead to disaster. The contrast in attitude to the two types of safety could hardly be starker.

Process safety indicators

If injury statistics provide no information about process safety performance, they must be supplemented with other measures that do. This was one of the strongest messages of the Baker Report. What might these process safety indicators look like?

Major accidents are rare and it is therefore not sensible to talk about a *rate* of major accidents at any one site. A site may go for years without a major accident. Accordingly, its annual major accident rate is zero, but it would be foolish to assume from this that the site is managing its major accident hazards well. Certainly this was not the case at Texas City.

The precursors to major accidents are, however, more common. Precursor events culminate in disaster only when certain other factors are present. A wide variety of precursor events can be identified, depending on the industry and the circumstances. If a site can drive down the number of such events, it can reasonably claim to be improving its process safety performance.

---

11   Willis, K, deposition, vol 1, 14 December 2005, p 34.

Some precursor events are obviously dangerous and therefore obvious candidates for use as process safety indicators. Loss of containment incidents, that is, accidental releases of flammable material, provide a good example. Interestingly, Texas City was already measuring loss of containment incidents at the time of the accident and the figure got steadily worse from 2002 to 2004, increasing by 52% in the two-year period, from 399 to 607.[12] This was a critical process safety indicator, but it was not a measure that mattered to BP; it was not a measure that commanded attention and it was not a measure that was used to drive improvement. Perhaps even more surprisingly, there were numerous fires at Texas City, but these were not seen as relevant to safety.

This led the Baker Panel to recommend that BP adopt a composite process safety indicator consisting of the number of fires, explosions, loss of containment events and process-related injuries.[13] The US Centre for Chemical Process Safety has subsequently recommended that the chemical industry as a whole adopt such a measure.[14]

Where a site is experiencing numerous fires and loss of containment incidents, as Texas City was, such a measure is a useful indicator of how well process safety is being managed, in the sense that a reduction in such incidents will imply an improvement in process safety management. At some sites, however, the number of fires and loss of containment incidents will already be so low that such figures cannot be used to monitor changes in the effectiveness of the process safety management system. An alternative approach is required in these circumstances.

This alternative approach begins by noting that the prevention of major accidents depends on a series of controls or barriers. Major accidents occur when all of these controls fail simultaneously. Where some but not all of the controls have failed, we can sometimes talk of a near miss or at least a degraded state of safety. From this point of view, the effectiveness of process safety management at a site can be assessed by evaluating the effectiveness of the controls that are supposed to be in place. To give a simple example: if safety depends on pressure relief valves opening when required, then what is needed is some measure of how well they are functioning. Or a different kind of example: if one of the controls on which safety depends is a requirement that operators stay within pre-determined process limits, then we need a measure of the extent to which they are exceeding those limits.[15]

---

12   Baker Report, p 187; CSB Report, p 172.

13   Baker Report, p 253.

14   Center for Chemical Process Safety, *Process safety leading and lagging metrics*, New York, CCPS, 20 December 2007, pp 5, 6. Available at www.aiche.org/ccps.

15   This whole approach is well discussed in Health and Safety Executive, *Developing process safety indicators*, London, HSE Books, 2006. I have deliberately avoided any reference to leading and lagging indicators in this discussion because, in my view, that distinction is confusing. See Hopkins, A, Thinking about process safety indicators, *Safety Science* (in press).

To return to Texas City, information on some of these issues was available on site, but it was not used to assess the effectiveness of the process safety management system. Certainly it was not used to drive improvements in that system.[16]

Triangle thinking

The assumption that injury statistics provide an indication of how well major hazards are being managed is widespread in the chemical industry. One of the theoretical supports for this misconception is the triangle (sometimes iceberg or pyramid) model of accident causation. It is worth addressing the problems with the model at this point.

The triangle model is based on empirical research by Bird and also by Heinrich.[17] These authors found that, for every serious injury that occurred at a site, there were a certain number of minor injuries, a larger number of property damage incidents without injury, and an even larger number of near misses. The triangle model assumes that these ratios are fixed in some way and that near misses and property damage cases are precursors to more serious injury, in the sense discussed earlier. The conclusion is that, by driving down the number of precursor events, the number of injuries and, in particular, serious injuries can be reduced. Figure 6.1 is a typical injury triangle.

Figure 6.1: Typical injury triangle

The original triangle theorists did not explicitly extend their triangles to cover major accidents, but others have implicitly done just that. In effect, it has been assumed that for every major accident there will be a certain number of serious

---

16   CSB Report, p 152.

17   Hale, A, Conditions of occurrence of major and minor accidents, *Institution of Occupational Safety and Health Journal* 2001, 5(1): 7–21.

injuries, a greater number of minor injuries, and so on. This implicit thinking is made explicit in Figure 6.2. The problem is now clear. There is an assumption here that the precursors of serious injury are also the precursors of major accidents. However, as we know, personal safety and process safety hazards are distinct and the precursor events for personal injury are, in general, not the same as the precursor events for major accidents.

Figure 6.2: Injury triangle extended to cover major accidents

Strictly speaking, every type of accident has distinctive precursor events, and every hazard requires its own triangle. But we can conveniently summarise the situation with two overlapping triangles, one for personal safety hazards and the other for process safety or major hazards (see Figure 6.3). The precursor events will, in general, be different, although there may be some in common — hence the overlap. It follows from the two-triangle model that, if we want to reduce the risk of major accidents, we must identify the precursor events that are specific to these accidents and set about reducing them in number.

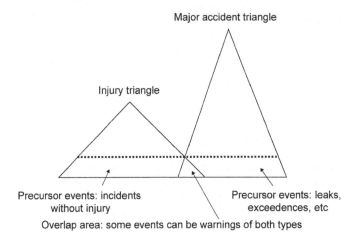

Figure 6.3: A two-triangle model

Figure 6.4 shows an attempt by one company in the oil and gas industry to give some substance to the major accident triangle in its context.[18]

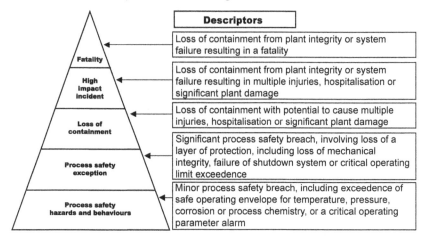

Figure 6.4: One attempt to specify a major accident triangle

The single-triangle model exercises a profound hold over much safety thinking and is both a cause and a symptom of blindness to major risk. The two-triangle model is therefore important. It provides a reconceptualisation that is far more consistent with the realities of major hazard facilities, and it focuses efforts appropriately.

The airline industry has led the way with two-triangle thinking. It recognises that flight safety and employee safety are two different things. For employees on the ground, such as baggage handlers, there is no overlap between the triangles at all. Many airlines maintain one database for near-miss incidents that affect flight safety and a separate database for matters affecting workforce health and safety. Moreover, they understand that workforce injury statistics tell us nothing about the risk of an aircraft crash. No airline in its right mind would seek to convince the travelling public of how safe it is by telling us about its workforce injury statistics.

The triangle model at Texas City

Figure 6.5 is an actual triangle diagram that was used in presentations at Texas City.[19] Let us identify the thinking that lies behind it. First, the top event is a fatality. There is an implicit assumption here that fatal events and more minor injuries are indistinguishable in terms of their causes, that luck plays a large part

---

18   See Appendix 2.

19   See presentation entitled *Texas City Refinery Safety Challenge*, 2001.

in determining whether, on any particular occasion, the outcome is one or the other, and that reducing the number of minor injuries will reduce the number of fatalities. We have already noted the flaw in this thinking.

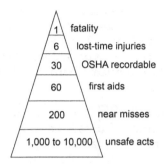

Figure 6.5: The Texas City triangle model

The base of this particular pyramid involves a second assumption, namely, that unsafe acts are the precursors to fatalities and that the risk of a fatality can be reduced by reducing the number of unsafe acts. This assumption provided the basis for a behavioural observation program at Texas City, in which workers were asked to make observations on each other and report any unsafe behaviour that they saw. This, it was hoped, would drive down the number of unsafe acts and so reduce the fatality risk.

It is important to note that behavioural observation programs encourage a focus on behaviour with certain characteristics. First, it must be readily observable. Second, it must occur frequently so that observers who devote a specific period of time to making observations can identify cases of compliance or non-compliance. These twin characteristics naturally highlight behaviour such as the wearing of PPE and the use of handrails when going up or down stairs.

Many types of unsafe behaviour are systematically missed by this approach, either because they are infrequent or because they are not obvious to the casual observer. Consider, first, the issue of frequency. One of the common causes of accidents is the short cuts that workers take in the attempt to restart a process that has temporarily jammed or broken down in some way. Workers may put their hands into a dangerous machine or crawl into a dangerous place, contrary to explicit safety rules, because they know that this is the most effective, perhaps the only, way to get the process restarted. Workers who in normal circumstances may be scrupulous about complying with rules (such as the wearing of PPE) will sometimes throw caution to the wind at times of crisis in order to get production going again. Unsafe behaviour of this nature does not fulfill a crucial characteristic mentioned above. It is relatively rare, precisely because breakdowns are abnormal, and casual observers are therefore not likely to be carrying out observations when it occurs.

The second problem is that some kinds of unsafe behaviour are not readily observable. Numerous accidents occur because of failures to properly implement permit to work systems, or to isolate plant that is supposed to be isolated, or to respond appropriately to alarms. The casual observer is not in a position to identify unsafe behaviour of this type. Indeed, there may be whole sequences of behaviour that need to be observed or even studied before it can be said that the behaviour in question is unsafe.

Consider, now, the failure to comply with startup procedures, such as occurred at Texas City. This is a form of unsafe behaviour that will almost certainly be missed by a behavioural observation program, both because startup is an unusual event and because the behaviour concerned is complex and not readily observable. The point is generalisable: most behavioural observation programs are unlikely to be of assistance in identifying process safety issues.[20]

In summary, the injury pyramid in use at Texas City, and the behavioural observation program with which it was associated, were systematically focused on personal safety, at the expense of process safety. They contributed to the blindness to major accident risk that was so evident at the refinery.

The failure to learn from previous process incidents

The lack of any focus on process safety indicators at Texas City meant that process safety incidents themselves appeared less significant. As a result, such incidents were not properly reported or analysed and lessons were not identified. Numerous incidents had occurred at the processing unit in which the explosion occurred. Had Texas City learned the lessons of these specific incidents, the explosion would never have happened. This section aims to provide some understanding of how dramatic these incidents were and how glaring the failure to learn.

In 14 of the 18 previous startups, operators had filled the distillation column beyond the range of the measuring instrument.[21] These were cases that should have been investigated and learnt from, but weren't. Nevertheless, they were not obviously dangerous incidents. In what follows, I shall restrict attention to incidents where the potential for escalation is obvious.[22]

---

20 For a more detailed discussion, see Hopkins, A, What are we to make of safe behaviour programs?, *Safety Science* 2006, 44: 583–597.

21 CSB Report, p 75.

22 The section draws on CSB Report, pp 307–312.

On six previous occasions in the preceding 10 years, there had been releases from the vent which produced vapour clouds at or near ground level. Instruments had recorded high vapour readings at ground level. In two cases, operators thought the event serious enough to shut down the unit. On four occasions, the matter had been serious enough to call firefighting staff, who responded in some cases by spraying the vapour cloud with fog to reduce the risk of ignition. These releases had been caused in a variety of ways. One was particularly notable: operators had filled to overflowing *another* distillation column that then discharged through the same vent. This was in many ways a dress rehearsal for the fatal explosion in 2005.[23]

In addition, on two occasions in the preceding 10 years, flammable material leaking through the pressure relief system had caught alight. In one case, the fire was extinguished by injecting steam into the system, and in the other case, operators tried unsuccessfully for three days to extinguish the fire, before deciding to shut down the production process to cut off the supply of fuel.

In summary, the process unit where the explosion took place had had at least eight dangerous occurrences in the preceding 10 years. They were dramatic events that required emergency interventions of various sorts. What had Texas City learnt from these events? The answer appears to be: very little. Neither of the two fires was recorded in any incident database and neither was investigated. Of the six ground-level vapour cloud events, only two were reported and investigated as safety issues. Four were reported as environmental issues. Where investigations were conducted, none questioned the basic design or system of operation that was leading to these events. Perhaps most tellingly, information about these eight events was not readily available to the CSB, which found it necessary to sift through a variety of sources in order to reconstruct this history of near misses.[24] It was not a history that was recognised at Texas City; it was not a history from which anything was learnt. Had this historical record been assembled and publicised, it would have been clear that it was only a matter of time before one of these incidents culminated in a major accident. Research shows that major accidents are always preceded by warning signs that, for one reason or another, are not recognised as such.[25] The warnings of danger in this case could hardly have been stronger.

The safety meeting described earlier is an example of how Texas City worked to keep personal safety issues in the forefront of people's minds. One can imagine similar staff meetings where stories about process safety near misses are told and retold, and workers are reminded of how these things can happen and, most importantly, how they can be avoided. A vapour cloud at ground level is surely a frightening event, a dramatic near miss. Reminding workers of previous

---

23   CSB Report, p 312.

24   CSB Report, p 310.

25   Turner, B, *Man-made disaster*, London, Wykeham, 1978.

vapour cloud events is one way in which awareness of major risk can be heightened. Near misses are often described as free lessons, but it takes organisational commitment to ensure that such lessons are remembered and embedded into operations. There was no such commitment at Texas City.

The failure of HAZOPs to consider previous incidents

There are various other, quite specific ways in which Texas City failed to learn from previous incidents, ways that contributed directly to the disaster. One of these has already been discussed — the management of change process in relation to trailer siting. The leader of that risk assessment process said at interview that, had his team been aware of the history of near misses at the nearby vent, they would probably not have agreed to site the trailer where they did.

Another instance of this failure to learn from previous incidents concerns a basic risk assessment procedure known as a hazard and operability study, or HAZOP for short. A number of HAZOPs had been performed on the unit where the explosion occurred but they failed to identify the risk of overfilling. It is important to understand *why* they failed. A HAZOP involves a team of knowledgeable people sitting down with a diagram of a piece of plant and thinking systematically about the ways that the process could go wrong: what would happen if the pressure here was too high or too low; the flow, too high or too low; the volume, too high or too low; and so on. Clearly, a HAZOP relies to some extent on the imagination of the people concerned and on their understanding of what can happen.

One of BP's most senior HAZOP experts explained at interview that the HAZOP teams would never have considered the possibility that operators would overfill the distillation column to the extent they did. "None of my peers that I have spoken with would have envisioned a similar circumstance to that which happened on [the day of the accident]." It would not have been a "credible scenario".[26] A HAZOP team would have assumed "an attentive alert operator [who would] intervene upon an alarm", he said.[27]

This is, of course, a problematic assumption. History tells us that operators frequently fail to comply with procedures for all sorts of reasons, and the possibility that the column might be filled to overflowing is not an unrealistic or unimaginable scenario. Had this scenario been envisaged, a HAZOP team might have recommended some piece of hardware to prevent its occurrence, such as a trip that would cut off supply to the distillation column.[28]

---

26 Broadribb, M, deposition, 15 February 2006, p 34.

27 Ibid, p 36.

28 Ibid, p 35.

The failure of earlier HAZOP teams to consider this possibility points to a fundamental problem with the way HAZOPs were conducted at Texas City. One way to imagine how things might go wrong is to examine how they have in fact gone wrong in the past. This means examining data on previous incidents. If HAZOP teams had been able to do this, they would have realised that operators systematically ignored alarms and overfilled columns, that there had apparently been previous incidents in which operators had filled columns to the point of overflow, and that there had been occasions when vapour clouds had formed at ground level. All this would have made the teams far more cautious in their assumptions about operator behaviour and far more likely to recommend a hardware upgrade to prevent such things from happening. But it seems that the HAZOP teams did not make use of data on previous process incidents in this way, perhaps because these data were not readily available. Their risk assessments therefore remained at a theoretical level, divorced, to some extent, from reality. BP's own analysis, after the event, is highly pertinent:

> "The 'What If' Analysis technique is not robust enough to consider all modes of operation or process upset scenarios. Nor are any reviews adequate if real incident and historical data is not considered. The ability to identify the major risks present on the site is highly dependent upon the level of risk awareness of the individuals involved in the process. The lack of reporting of process excursions/incidents also hinders all types of hazard analysis."[29]

The major lesson: the need to focus on process safety

One of the main conclusions drawn in all of the reports arising from the Texas City accident was that BP had focused on personal safety at the expense of process safety. All noted, in particular, that BP had emphasised personal safety indicators and had paid little attention to process safety indicators. Correspondingly, their recommendations focused on the need to give greater prominence to process safety.

Some of the strongest statements to this effect were contained in the Baker Report on BP's five North American refineries, as demonstrated in the following passage:

> "The Panel believes that BP has not provided effective process safety leadership and has not adequately established process safety as a core value across all its five US refineries ... BP has emphasized personal safety in recent years and has achieved significant improvement in personal safety performance, but BP did not emphasize process safety. BP mistakenly interpreted improving personal injury rates as an indication of acceptable process safety performance."[30]

---

29   *Fatal Accident Investigation Report*, London, BP, 9 December 2005, p 67.

30   Baker Report, p xii.

Accordingly, *every one* of the 10 Baker Panel recommendations refers explicitly to the need to focus on process safety. There could be no doubt about the single most important lesson arising out of the Texas City accident.

Chapter 7

Inability to learn

Major accident inquiries sometimes seek to identify lessons, so as to ensure that the type of accident under investigation never happens again. They implicitly assume that their role is to uncover new knowledge and identify lessons not previously available.

Unfortunately, however, accidents tend to repeat themselves. Perhaps the best known recent example of this repetition is the way that the Columbia space shuttle accident replicated the Challenger accident 17 years earlier. In both cases, an entire crew perished. The particular details of these accidents were different, but in both cases an underlying cause was the normalisation of risk: certain things regularly failed to function as intended, but they never failed totally, giving rise to a view that the partial malfunction could be considered normal. This process of normalisation was highlighted as a major lesson from the first disaster and the same pattern was identified in the second. The National Aeronautics and Space Administration (NASA) appeared to have learnt nothing from the Challenger accident. So striking was this similarity that a whole chapter of the Columbia report was devoted to discussing it.[1]

The Baker Panel explicitly saw its function as identifying the lessons from Texas City. Its final recommendation was that:

> "BP should use the lessons learnt from the Texas City tragedy and from the Panel's report to transform the company into a recognized industry leader in process safety management."

The fact is, however, that these lessons were widely available previously. In particular, there were numerous accident analyses stressing the need to focus on process safety and warning that injury rates reveal nothing about how well process safety is being managed. Furthermore, these lessons were not floating in the ether; they had been brought directly to the attention of people at Texas City.

This raises in an acute way the question of why these lessons had not been effectively learnt. In the previous chapter, we saw that Texas City did not learn from the process incidents occurring at the site because of a general lack of focus on process safety. In this chapter, I want to argue that the failure to focus on

---

1   Columbia Accident Investigation Board, *The Report*, vol 1, Washington, National Aeronautics and Space Administration, August 2003.

process safety involved such a serious failure to learn lessons already available that Texas City can be said to have suffered from some kind of learning disability.[2] Before drawing that conclusion, however, I need to demonstrate the extent to which the lessons about process safety had already been drawn to the attention of people at Texas City.

Grangemouth

Grangemouth, in Scotland, is one of the largest petrochemical facilities in Europe. Over a two-week period in 2000, it experienced three major incidents. The first involved a complete power failure and occurred when an underground power cable was ruptured by digging. In the second incident, an 18-inch steam line ruptured, causing a jet of steam that damaged fencing and escaped across a nearby public road. The escaping steam produced a roar that could be heard in the town of Grangemouth. Only one member of the public was injured, but there was an obvious potential for a much more serious outcome. In the third incident, a major leak from a processing unit ignited, causing a large fire that did considerable damage. No one was injured, but in slightly different circumstances, the leak could have resulted in a vapour cloud explosion with multiple fatalities.

The second and third incidents caused commotion in the town of Grangemouth and were reported in the national press. It was obvious that the site had been spared disaster by good luck alone. The occurrence of such a rapid succession of events triggered a major investigation by the regulator into the safety culture of the site, and a prosecution was launched that resulted in fines of £1m.

The owner of the Grangemouth site was none other than BP, which remained the owner until shortly after the Texas City explosion in 2005, when the facility was sold to another company. There is thus a direct organisational connection between Grangemouth and Texas City.

BP ordered its own inquiry into the Grangemouth events and assembled a task force to carry out a "root and branch" audit of the site. This was the largest such task force ever assembled by BP. It consisted of 36 engineers and safety experts. The BP inquiry came to broadly the same conclusions as the regulatory inquiry. Its findings were communicated widely throughout the BP empire and a "lessons learnt" workshop was held at Grangemouth with representatives from all of BP's business groups.

---

2   I take this idea of organisational learning disability from Barram, M, Shame, blame and liability: why safety management suffers organisational learning disabilities. In Hale, A, et al, *After the event: from accident to organisational learning*, Oxford, Pergamon, 1997, pp 163–178.

What *were* the findings? They were set out in various documents, with stark clarity, as a series of lessons and messages. Some of these are quoted below:

"Lesson 1: . . . Control of major accident hazards requires a specific focus on process safety management over and above conventional safety management."

"Lesson 2: Companies should develop key performance indicators (KPIs) for major hazards and ensure process safety performance is monitored and reported against these parameters."

"Message 1: Major hazard industries should ensure that the knowledge available from previous incidents both within their own organisation and externally is incorporated into current safety management systems."[3]

These lessons/messages are exactly the key lessons/messages arising out of the various reports on the Texas City accident. It seems we are doomed to be taught the same lessons again and again.[4]

We can reasonably conclude that, had the lessons from Grangemouth been properly assimilated at BP's North American refineries, the Texas City accident would never have happened. But is it reasonable to assume that the Texas City site was aware of lessons from across the Atlantic, even if both sites belonged to the same company?

The answer is that the lessons from Grangemouth were available to Texas City in a very direct way. First, BP's Grangemouth investigation was not a local matter; it was headed by a senior manager from the United States. There was an even more direct link. Three BP employees subsequently wrote and published an article spelling out the lessons from Grangemouth.[5] The authors' positions were:

- process safety specialist at Grangemouth
- process safety expert reporting to BP's top US executives, and
- process safety manager at Texas City.

The third author was a direct and personal conduit of information from Grangemouth to Texas City. The lessons from Grangemouth were not simply available to Texas City in the same way that they were available to other refineries around the world; they had a particular champion at the Texas City site. It is this that makes the failure at Texas City to learn the lessons from Grangemouth so remarkable.

---

3 Health and Safety Executive, *Major Incident Investigation Report, BP Grangemouth Scotland: 29th May–10th June 2000*, HSE and Scottish Environment Protection Agency, 18 August 2003, pp 74–76.

4 George Santayana once wrote: "Those who cannot remember the past are condemned to repeat it." *Flux and constancy in human nature*, New York, Charles Scribner's Sons, 1905, p 82.

5 Broadribb, M, Ralph, W and Macnaughton, N, Lessons from Grangemouth — a case history, *Loss Prevention Bulletin* 2004, 180: 18–26.

Longford

A second, quite striking example of the failure of Texas City to profit from earlier lessons is provided by the Esso Longford gas plant explosion near Melbourne in 1998. My book, entitled *Lessons from Longford*, was published in 2000. Within weeks, someone had produced a 19-page document of excerpts from the book, which circulated around the petroleum industry via email. The document quoted verbatim the following lessons from the book:

● reliance on lost-time injury data in major hazard industries is itself a major hazard

● systematic hazard identification is vital for accident prevention

● corporate headquarters should maintain safety departments which can exercise effective control over the management of major hazards

● frontline operators must be provided with appropriate supervision and backup from technical experts

● routine reporting systems must highlight safety-critical information

● maintenance cutbacks foreshadow trouble, and

● companies should apply the lessons from other disasters.

These are all lessons that Texas City needed to learn. And again, they were not merely available to Texas City; they were actively brought to the attention of management. *Excerpts from Lessons from Longford* was circulated within BP and specifically at Texas City on various occasions, sometimes with a covering email urging people to read the document.

One such chain was begun on 3 October 2000 by John Mogford, subsequently the leader of BP's Texas City investigation team. His covering email said:

"A copy of a report that I read on a major Exxon incident in Australia is a long tome but worth a scan, some of the cultural/corporate issues sound familiar."

One of his recipients passes it on with the following comment:

"Elaine, please print this out for me.

Mike and Bill, I've heard from a variety of sources (including the head of HSE for Petronas) that this is an excellent report. You may want to skim it."

It turns out that Bill is the director of HSE (health, safety and environment) for BP Chemicals in the US. He passes *Excerpts from Lessons from Longford* on with the comment:

"I highly recommend the attached reading. It is very thought provoking and has me looking at our HSE programs with a different lens. Print it out and take some time to read the 19 pages."

One of his recipients passes it on with this comment:

> "Read this report!! Some very interesting learnings that hit close to home. As an added bit of interest, BP developed gHSEr from Exxon's OIMS system.[6,7] Should we do anything different as we move forward in PF and our Change Management efforts? As noted below, best if you print out and read . . ."

It is clear from these comments that the lessons from Longford struck quite a chord with those who read *Excerpts from Lessons from Longford*. They realised that BP needed to take a new approach: major hazards required a particular focus and could not be managed in the way that conventional hazards were.

There is one point about this email chain that cannot pass without comment. The senders realise that the receivers will be disinclined to read the document because of its length (it's "a long tome but worth a scan"). Several senders urge receivers to print it out, presumably thinking that it is more likely to be read in the printed form. Managers, it seems, have little time for reading. This is, no doubt, one of the factors undermining their capacity to assimilate lessons from elsewhere.

The Longford example was also used in training courses at Texas City. The instructions to participants were as follows:

> "Please read (or re-read) the attached excerpts from Lessons from Longford. The reading contains powerful insights into thinking about integrity management and safety that are quite different from how we usually think about these subjects."

Participants were asked to reflect on recent incidents at Texas City in the light of the Longford material.

Finally, the process safety manager at Texas City had detailed knowledge of the lessons from Longford and drew on them in his attempts to influence management.[8] In one email, he wrote:

> "A strong and successful personal safety program does not guarantee a strong and successful process safety program. Esso Australia learned this lesson."

In summary, the lessons from Longford were frequently and forcibly brought to the attention of management at Texas City, but apparently they failed to have the necessary impact.

---

6   Getting Health Safety and Environment Right (gHSEr).

7   Operations Integrity Management System (OIMS).

8   Ralph, W, deposition, vol 2, 28 July 2006, p 43.

Messages from major reviews

It was not just that Texas City seemed incapable of learning lessons from elsewhere. It also seemed unable to respond to some very explicit messages provided by high-powered review teams. A major external review of Texas City in 2002 described the failings that it identified as "urgent and far-reaching".[9] It went on:

> "Asset safety [process safety] . . . is one of the biggest issues identified by the assessment team. While personal safety performance at South Houston is excellent, there were serious concerns about the potential for a major site incident due mainly to the very large number of hydrocarbon escapes [over 80 in the 2000–2001 period]. Also there is a large backlog of overdue inspections . . ."

In 2003, a major audit was carried out by a high-powered team of BP people who were external to Texas City. It was called a "getting HSE right" audit, and was not designed to focus on process safety. Nevertheless, it concluded that the leadership team had not done enough to address process safety risks.

Finally, a survey conducted just a few months before the explosion found that the workforce itself believed that management worried too much about seat belts (personal safety) and too little about catastrophic risk.[10] The authors of the report commented that they had never seen such "intensity of worry" about the possibility of disaster by those "closest to the valve". "This has had a profound impact on us all", they said. "We have never seen a site where the notion 'I could die today' was so real for so many . . . people." "There is an exceptional degree of fear of catastrophic incidents at Texas City." The report noted that concerns were not restricted to frontline workers. "The more expert interviewees [engineers, inspectors, etc] voiced the same strong concern, as did superintendents and certain leadership team members." Contractors who were interviewed said that the situation was worse than they had experienced at Shell, Chevron and Exxon. The results of this survey were not effectively transmitted to more senior BP people. Even if they had been, the warning was probably too late to make a difference.

---

9   *Good practice sharing assessment* (the Veba Report), August 2002, pp 4, 9, 75, 76, 105. Available at www.texascityexplosion.com.

10   The Telos Group, *BP Texas City site report of findings,* 21 January 2005. The quotations in the text are taken from three documents produced in the course of this review: (1) a PowerPoint presentation entitled *Texas City site integrity and safety leadership, update to leadership team,* 13 December 2004; (2) *Executive summary of report of findings,* 21 January 2005; and (3) a memo entitled "What makes protection particularly difficult for BP Texas City?". This memo is referred to in *Investigation Report: Refinery Explosion and Fire* (CSB Report), Washington, US Chemical Safety and Hazard Investigation Board, March 2007, p 175. All of the Telos documents are available at www.texascityexplosion.com.

It is worth noting that the Texas City HSE manager, in particular, shared the fears that were so widely expressed by the workforce. Almost exactly a month prior to the explosion, he wrote in an email to the leadership team at the refinery:

"I would like for us to make these incidents our No 1 priority (at tomorrow's meeting). I truly believe that we are on the verge of something bigger happening and that we must make critical decisions tomorrow morning over getting the workforce's attention around safety."[11]

There is something of a paradox here that needs to be addressed. When tracing the accident sequence in earlier chapters, we uncovered many examples of a lack of awareness of explosion risk among employees at several levels. This led me to speak of a blindness to major risk. Yet the evidence from the workforce survey suggested that people were deeply worried about the possibility of disaster. How can these opposing findings be reconciled?

The context of the survey is important. It was carried out a few weeks after a tragic incident that occurred in September 2004. Two employees were killed and another was seriously injured when burned with hot water and steam during the opening of a pipe flange.[12] Of course, this created a sense of fear among people working at the site. Moreover, this was arguably a process safety matter. But it did not open people's eyes to the need for a different approach when it came to process safety. So it was that the realisation that "I could die today" went hand in hand with a lack of any specific focus on process safety risks. The widespread sense of foreboding was not matched by any recognition of what could or should be done to reduce such risks.

The reviews and audits in the years prior to the explosion set the Texas City case apart somewhat from other major accidents. Prior to the Longford accident, a major audit carried out by the parent company, Exxon, had given Esso Australia a clean bill of health. This had the effect of lulling management into a false sense of security. The inquiry into the Longford accident was extremely critical of the audit team's failure to identify all of the problems that became obvious after the accident. Very similar comments can be made about the world's worst petroleum accident on the Piper Alpha platform, off the coast of Scotland, in 1988.[13] The inquiry into that disaster also noted that prior audits had provided a false sense of security. The same cannot be said in the Texas City case. Various audits and reviews had provided very clear messages about the dangerous state of the plant, yet Texas City management seemed incapable of responding effectively.

---

11   Email dated 20 February 2005.

12   CSB Report, p 220.

13   Hopkins, A, *Lessons from Longford: the Esso gas plant explosion*, Sydney, CCH Australia Limited, 2000, pp 80–84.

A learning disability?

The failure to learn the lessons and to heed the messages is so striking that it is not unreasonable to conclude that Texas City suffered from some kind of learning disability. The term is dramatic, but so is the failure to which it refers. I do not wish to suggest that Texas City was unique in this respect; the term is equally applicable to NASA in relation to the space shuttle accidents, and it is probably applicable in many other contexts where accidents repeat themselves.

To speak of an organisational "learning disability" draws attention to the need for explanation. Many accident investigations aim to identify lessons to be learnt. The question of interest at this point is why organisations fail to take account of lessons that are *already available*. What stops them from implementing these lessons? Texas City is an excellent case study from this point of view.

We are now at a turning point in this book. Previous chapters focused on the direct causes of the accident. The next four chapters will examine higher-level organisational causes that contributed to the failure to implement lessons from elsewhere. We have already noted one such higher-level cause: the fact that process safety was dealt with as a matter of risk management, not rule compliance, which meant that it was systematically de-emphasised in BP's budgeting priorities. Four other reasons will be identified in following chapters: cost cutting, an inappropriately focused incentive system, decentralisation, and a lack of process safety leadership.

Chapter 8
Cost cutting

One of the reasons for the failure to respond to lessons or even to heed warnings was that Texas City had been paralysed by years of cost cutting. A review for BP in 2002 put it a little more tactfully: "The current integrity and reliability issues at Texas City Refinery are clearly linked to the reduction in maintenance spend of the last decade."[1] The Chemical Safety and Hazard Investigation Board (CSB) agreed: "Cost cutting, failure to invest and production pressures from BP Group executive managers impaired process safety performance at Texas City." Eva Rowe made a far more dramatic claim: ". . . the explosion was caused by [BP's] pure, unadulterated GREED!"[2] In this chapter, I want to explore how this cost cutting came about.

There is something almost paradoxical about the pressure that Texas City was under to cut costs, because the refinery was very profitable. The year prior to the accident was the most profitable in its history: it made $930m.[3] Moreover, this was $145m more than any other refinery in the BP group. It was indeed BP's flagship. As one lawyer said, Texas City was a cash cow for BP.[4]

Capital costs

Profit is not, however, the main story. The problem was that there was a great deal of capital invested in Texas City and the return on that investment had generally been below 10%.[5,6] BP saw this as insufficient.

Some readers will be unfamiliar with the concept of return on investment, so an example may help. If an asset such as a processing plant is valued at $100m, and the asset is generating an annual profit of $10m through its processing activities, then the annual return on the capital tied up in the plant is 10%. If owners are aware of a second asset that is generating a higher return, it is in their interest to sell the first asset and buy into the second.

---

1 *Texas City Refinery retrospective analysis*, October 2002, p 11. Unless otherwise indicated, the documents referred to in this chapter are unpublished internal documents, available at www.texascityexplosion.com.

2 Statement to the media, 4 February 2008 (emphasis in original). See also statement to the media, Houston, 4 February 2008: "BP's greed murdered 15 people!"

3 Replacement cost operating profit. Figures from Parus, D, update, March 2005.

4 Parus, D, deposition, vol 1, 22 June 2006, p 11.

5 Return on average capital employed.

6 *Texas City Refinery retrospective analysis*, op cit, p 7; *TCR Refining Leadership Training Away Day*, 21 August 2003, slide 18.

An asset tends to run down over time, and an annual injection of capital is necessary in order to maintain it in good operating condition. This cuts into profit. It follows that one way to increase the return on investment, at least in the short run, is to decrease the annual capital injection. Accordingly, capital expenditure had been steadily reduced by BP, and by Amoco before it.[7] In particular, capital expenditure for safety and for maintenance had been substantially reduced. The 2002 review put it very simply: "... capital [expenditure] was reduced in order to improve returns."[8]

Even so, a 2003 analysis suggested that annual capital expenditure needed to be reduced further. Among BP's 18 refineries worldwide, Texas City was seen as requiring more than its fair share of annual capital expenditure.[9] In a memo at the time to all refinery employees, a senior manager wrote the following:

"The Texas City Refinery due to its size, complexity and environmental requirements, has a significant appetite for capital dedicated to maintaining a safe, productive and compliant facility. As a result it is projected to consume 18% of the capital [available for all BP refineries] over the next five years. This in and of itself would not be a problem but unfortunately the profit contribution for Texas City over that same time period is projected to be only 15% of the total for all the refineries. Clearly the corporation can not continue to invest money in Texas City when it is out of proportion to what the Refinery can generate in profits relative to our other refineries."[10]

The memo went on to outline what it described as "sobering" information:

"Though Texas City is the largest consumer of capital in the refining [group], benchmarking data suggests that we have invested less on our facility than much of the competition over the last several years.

Though we are one of the most complex refineries in the world, many of our US Gulf Coast competitors are able to extract more value from their less complex facilities ..."

It was the implication of this that was most sobering. BP was already underinvesting in Texas City, relative to the competition. To put it bluntly, the site was being allowed to run down. The deterioration was evident to all, both in the physical appearance of equipment and in terms of various indicators of performance.[11] But despite this underinvestment, returns remained unsatisfactory. Further cuts in capital expenditure were unlikely to produce the desired result and the only way forward was a radical one: to close down some

---

7   *Texas City Refinery retrospective analysis*, op cit, p 7.

8   Ibid.

9   *Refining and pipeline SPU Texas City Refinery review*, bilateral with John Manzoni, 24 November 2003.

10   Hale, R, memo, 14 November 2003.

11   *Texas City Refinery retrospective analysis*, op cit, p 7.

of the units at the site that were requiring the greatest capital expenditure to keep them going. Accordingly, the memo warned that:

"[Texas City will] examine options to reconfigure the refinery [that is, close down some units] to increase its competitiveness and reduce the capital spend."

A covering email summarised the memo as follows:

"The key message is Texas City Refinery is NOT delivering on profitabilty vs % of capital investment. This is part of our 'sense of urgency'."

But the logic of "reconfiguration" was not followed through. In 2004, BP "challenged" its North American refineries to cut their capital expenditure estimates for the following year by 25%.[12] The cuts were to be achieved by deferring projects of various sorts, rather than by closing down the units that required disproportionate amounts of capital. The Texas City site manager argued against the 25% cut and it was reduced in his case to about 16%. But this cut was then imposed. This was not a challenge that the manager could choose to ignore.

It is possible that even closing down some of the most capital-expensive processing units at Texas City may not have solved the problem. One of BP North America's top executives, who by his own admission recognised the impact that underinvestment was having at the site, argued at one stage that BP should sell off refineries that it could not maintain. However, he did not follow through with this in the case of Texas City, and the result was that the policy of underinvestment continued.[13]

A note on language at this point. The memo above states that Texas City had "a significant appetite for capital dedicated to maintaining a safe, productive and compliant facility". The phrase "appetite for capital" is set in a context in this case, but time and again in discussions of Texas City costs, the phrase occurs by itself, that is, without reference to why the capital is needed. In these circumstances, the metaphor takes over. It is as if Texas City is some kind of monster that is devouring capital and, although this particular monster cannot be slain, it must at least be constrained. The use of the metaphor subtly undermines requests for additional capital.

---

12 Parus, op cit, pp 45, 57. See also Dio, S, email, 27 August 2004.

13 *Management accountability project* (Bonse Report), February 2007, p 10; Hoffman, M, interview for Bonse accountability project, 14 June 2006, p 2. It was indeed a policy of underinvestment. In 2003, BP issued a document called a "management framework" that stated: "While acknowledging that the refining segment had grown considerably in the past few years due to acquisitions and that it could generate high returns, a sub-strategy was to limit the amount of capital allocated in the Refining SPU due to its 'volatility'" (*Investigation Report: Refinery Explosion and Fire* (CSB Report), Washington, US Chemical Safety and Hazard Investigation Board, March 2007, p 152).

Operating costs

A second way that return on investment can be increased is by cutting the annual operating costs, for instance, by cutting staffing levels, training budgets, and so on. Texas City was also under enormous pressure to cut such costs. It began as soon as BP had acquired the site from Amoco in 1999. BP's CEO set new business goals for the whole of BP, one of which was to cut cash costs by 25%. Amoco had already been cutting costs at Texas City for several years, and the new cuts were to be over and above those already achieved. Nevertheless, Texas City dutifully developed a business strategy to achieve the new goal. A central feature of the strategy was to lay off staff. The goal was 15% fewer employees. This would be achieved by "reducing layers and staffing in the organisation through a severance program", the strategy stated.

An assumption here is that it is essentially wasteful to have several layers of staff, and that flatter organisations with fewer layers are more efficient. However, the flatter an organisation, the greater the span of control for any one manager, and the wider the range of activities that must be managed. It is clear that if too many layers are removed, managers will be unable to manage effectively. Evidently that is what happened at Texas City for, as we saw, the frontline operators and their immediate supervisors were left very much to themselves, with little or no supervision from more senior staff.

Another way that Texas City aimed to achieve the 25% cost reduction was by cutting various training programs. Finally, numerous routine maintenance jobs were deferred.[14] The safety implications of these cuts are obvious.

The strategy stressed that:

> "The Texas City Business Unit will continuously and aggressively drive costs out of the system at an accelerated pace relative to other refiners."

The image of "continuously and aggressively driving costs out of the system" is arresting. It occurs twice in the strategy statement, suggesting that it is significant in the mind of the document's author. Its implications are disturbing. It implies that costs are not integral to production but are foreign bodies that have somehow found their way, uninvited, into the system. They are to be eliminated. There is no recognition that these costs may have been incurred for legitimate purposes. As in the case of "appetite for capital", the image obscures important issues and biases decisions in a particular direction.

Those responsible for making the cuts at Texas City doubted whether they were sustainable, that is, they were concerned that there would need to be compensatory budget increases in later years.[15] But they made the cuts anyway, believing that they had no option.

---

14   CSB Report, p 99. Various other documents giving details of the cuts are available at www.texascityexplosion.com.

15   Maslin, P, deposition, 13 July 2006, p 10.

The 1999 cost challenge was not a one-off occurrence. In 2002, BP's second-in-command issued a memo to all refinery leaders urging them, among other things, to challenge all costs wherever possible, to defer training programs to the New Year, and to freeze all hiring.[16] Clearly, the costcutting pressure was relentless.

How did other BP refineries respond to the 25% costcutting challenge?

As we saw, management at Texas City took the 1999 challenge as an instruction. As one later review put it: "... the prevailing culture at the Texas City Refinery was to accept cost reductions without challenge and not to raise concerns when operational integrity was compromised."[17]

By 2002, senior BP executives were musing about how Texas City had "gotten into such a poor state". One of them noted that the 25% cut "seems to have been taken literally at [Texas City] whereas [the manager at Whiting, another North American refinery] knew how to play the game".[18] It is apparent from this comment that the manager at Whiting did not cut costs by 25%, as challenged to do.

The comment also suggests that BP head office in London did not necessarily expect its managers to achieve a full 25% cut. However, in the same document, the challenges are referred to as orders and the view is expressed that the top level in London did not understand the consequences of its orders.

There is considerable debate over the extent to which these cost challenges amounted to instructions. The manager at Whiting was able to deflect the challenge with impunity. Others were not so lucky.[19] Refinery managers in the United Kingdom were told that the challenge was a "directive". One of these managers, at Coryton Refinery, told his superior that he "didn't think it was realistic to achieve a 25 per cent cost reduction". He had already been cutting costs over a number of years and he was unwilling to commit to a further 25%. His superior had not appreciated his comment, he said later; it created tension between them. Subsequently, his position was changed in such a way that he was no longer responsible for business decisions, although he remained the refinery manager. This was effectively a demotion. He was being punished for speaking out against the 25% cut, he said. But he was not sorry he had done so; the 25% cut was neither safe nor sensible. The manager said he was not aware of any other refinery managers making the stand that he had made.

---

16 Manzoni, J, memo, 23 October 2002.

17 *Texas City Refinery retrospective analysis*, op cit, p 10.

18 Email.

19 Maslin, op cit, pp 26, 80, 81.

There is no doubt that the pressure perceived by the manager at Coryton was perceived at Texas City as well. A memo written by a senior manager at Texas City, by chance just hours before the explosion, demonstrates how powerful these pressures were:

> "We are over-running our entire [maintenance] budget tremendously, 20 million plus, but I have to say that I couldn't turn some of this work down as it is critical to the safety of the unit, corrosion issues, cracking metal, etc. I'll probably ultimately get fired over some of the cost issues, but I have to feel I am doing the right thing."[20]

As at Coryton, this manager was resisting the cost pressures, but she believed that she might pay for this with her job — in much the same way that the Coryton manager had. While London head office may at times have suggested that managers were within their rights to resist cost challenges, many managers on the ground believed that they did so at their peril.

Benchmarking

It is clear from the earlier discussion that spending decisions were not made on the basis of what was necessary to keep the plant operating reliably and safely, but on the basis of considerations such as what the competitors were spending. Recall the statement that Texas City would "drive costs out of the system at an accelerated pace *relative to other refiners*". This phenomenon of comparing oneself with the competition is described in the industry as benchmarking. The industry average is taken as a benchmark and the aim is to increase returns, or reduce spending, or improve reliability, or cut staffing levels, so as to at least equal, and preferably surpass, industry averages.

A company by the name of Solomon Associates collects a wide variety of data in the oil and gas industry and presents the data back to the industry in such a way as to facilitate this benchmarking.[21] References to Solomon benchmarking are everywhere in the documentation on the Texas City accident. Senior executives admit that they "made extensive use" of Solomon benchmarks and that they were engaged in "continuous conversations" about costs in which Solomon indicators figured prominently.[22] Solomon data were used in 1994 to argue for an 18% cut in staff at Texas City.[23]

The Solomon data enable industry players to identify which quartile they lie in with respect to each of the indicators, and the aim expressed by many is to occupy the top quartile. Of course, by definition, only a quarter of players can be in the top quartile, so the result of this benchmarking activity is a constant

---

20   Maclean, C, deposition, 12 September 2006, p 32.

21   CSB Report, p 154.

22   Bonse-Geuking, W, deposition, vol 2, 14 February 2007, pp 8, 41. See also Wundrow, W, deposition, vol 2, 29 August 2006, pp 19, 20.

23   CSB Report, pp 88, 89.

struggle to cut costs, staffing, and so on, without regard to the circumstances of a particular plant. Texas City Refinery was one of the most complex in the world and required relatively large amounts of capital to keep it going. There was good reason, in other words, why it might find itself in the bottom quartile and why its owners should have been content with bottom quartile performance. But senior executives did not accept this situation and they expected Texas City to make cuts, without fully understanding the consequences.

Solomon benchmarking can sometimes have ironic consequences.[24] One of the benchmark categories is reliability, defined as percentage of time that a process unit is up and running. If a unit is shut down for routine maintenance, this reduces its up time and, according to the definition, its reliability. Hence, reliability, as defined by Solomon, can be increased by delaying routine maintenance shutdowns. The irony is that delaying maintenance in this way increases the risk of breakdown, which can result in much longer periods of downtime. In short, improving Solomon reliability in this way decreases the real reliability of a unit.

All of this led the 2002 review to conclude:

> "Future maintenance spend optimization should be based on implementation of best practices in maintenance management ... and should not be driven solely by external benchmarks or across-the-board cost-cutting initiatives."[25]

The learning and development department

The impact of this cost cutting is clearly illustrated by the experience of Texas City's learning and development department, which was responsible for training. Between 1998 and 2004, the budget for this department was halved and its staff was reduced from 28 to eight.[26] To make up for this, Texas City had gone to computer-based training. This was explicitly a cost-reduction strategy, without consideration of the effectiveness of such training.

Computer-based training was widely seen as ineffective. Tests could be taken several times and workers learnt how to manipulate this by taking a test, seeing the correct answers, and then retaking the test. Sometimes supervisors took the tests on behalf of workers.[27] But the limitations were more fundamental than this.

---

24   Bonse Report, op cit, pp 21, 22. See also Wundrow, op cit, p 9.

25   *Texas City Refinery retrospective analysis*, op cit, p 11.

26   CSB Report, p 98.

27   *The Report of the BP US Refineries Independent Safety Review Panel*, Washington, US Chemical Safety and Hazard Investigation Board, January 2007, p 110.

As the CSB pointed out:

> "... operators who require training for abnormal conditions would not benefit from computer-based training that often focuses on memorizing facts, not troubleshooting unusual events."[28]

ust a week before the explosion, the Texas City manager acknowledged the inadequacy of existing training when he announced a new training initiative:

> "The training will be face-to-face versus computerised so that we can get the discussion that is needed to really achieve full understanding."[29]

The best form of training for unusual events, such as the situation which confronted the Texas City operators on the day of the accident, is simulator training. This requires the construction of simulated control rooms in which operators can experience a variety of abnormal situations in a short space of time. Airline pilots undergo simulator training in which they can crash the aircraft and learn from the experience, and air traffic controllers practise in a simulated environment where their mistakes do not put the lives of passengers at risk. Simulators have made their appearance in certain oil refineries and gas processing plants, but were not in use at Texas City. Starting in 2000, the head of the learning and development department had, on various occasions, urged Texas City management to invest in simulators. However, there was always a "big pushback", he said, on the grounds of cost.[30]

One can readily imagine exercises for refinery operators that simulate the overfilling of distillation columns. Given the prevalence of such events in the past, this would have to have been a high priority simulation. If the Texas City operators on the day of the accident had already experienced these events in a simulator, they would probably have recognised the danger a lot earlier and reacted in such a way as to avoid the accident.

BP has learnt its lesson in this respect and has decided to invest in simulators. In the words of the man appointed to manage Texas City after the accident: "[There will be] simulation training for each and every operator off the unit, so that they are experienced in dealing with upsets before they ever touch the controls."[31]

It seems fair to say that, had BP been willing to invest in simulators when they were first proposed, the accident would probably not have occurred.[32]

---

28 CSB Report, p 98.

29 Parus, D, email, 18 March 2005.

30 CSB Report, pp 97, 98.

31 Maclean, op cit, p 33.

32 BP's own investigators were clearly tempted to come to the same conclusion, but drew back on the grounds that "it is very difficult to quantify 'incidents that won't occur'" (*Fatal Accident Investigation Report*, London, BP, 9 December 2005, p 88).

Concluding comments

The impact of cost pressures on process safety is particularly clear in the preceding discussion. But the claim in this chapter is not just that the cost cutting affected training; it is that it undermined the capacity of the organisation itself to learn, that is, to respond to safety-critical information. There were numerous audits and reviews in the years preceding the explosion that identified training shortfalls, maintenance issues and supervisory inadequacies, all of which had a direct bearing on process safety. Moreover, there were plenty of lessons from elsewhere about the importance of process safety. To respond effectively to these messages would have required a significant increase in funding. This was impossible in an environment where cost cuts were regularly imposed on site managers by the top echelons of the company. In summary, cost cutting at Texas City was a major contributor to the failure to respond to lessons or to heed explicit warnings.

BP's position is that cost cutting was not a cause of the Texas City accident. "We can't find any *direct* linkage" (emphasis added), said Tony Hayward, who became BP's CEO two years after the accident. "Everything that we can find suggests that the budget cuts *per se* did not contribute to . . . the tragedy at Texas City" (emphasis added).[33] But a few months earlier, he had said:

> "The frontline operations teams, I think, have lived too long in the world of making do and patching up this quarter for the next quarter, rather than really thinking about how we are going to maintain a piece of equipment for the next 30 or 40 years . . . [BP's] mantra of 'more for less' . . . needs to be deployed with great judgment and wisdom . . . When it isn't, you run into trouble."[34]

These latter comments seem implicitly to recognise that budget cuts, while perhaps not a *direct* cause, were an underlying or root cause of the accident.

BP's cost challenges had this characteristic: senior executives demanded cost cuts and left it to others further down the hierarchy to ensure that these cuts were not at the expense of safety. Lower-level managers responded as best they could to these conflicting requirements but, inevitably, safety was compromised. The only way out of this predicament is for those who order the cost cuts to take responsibility themselves for ensuring that safety is not compromised. How might they do this? Obviously, the CEO cannot personally investigate staffing and maintenance needs at each site to ensure that cost cutting is not at the expense of safety. But he or she might appoint a team, directly answerable to the CEO, to go into the field and find out. The team would need to be independent of the cost pressures affecting line managers and, provided this was assured, the CEO could expect unbiased advice about the impact of cost cuts. Rather than

---

33 *Houston Chronicle*, 15 June 2007.

34 BBC News, 18 December 2006.

saying, "I rely on others to ensure that safety is not compromised", the CEO needs to be able to say, "I am personally satisfied that the impact of cost cuts has not been at the expense of safety". If an independent process of verification on behalf of the CEO had been in place, BP's top people might have become aware of the impact of their cost cuts before it was too late.

There is another way of making this point. BP's policies required sites to carry out risk assessments before making changes. This was not restricted to technical changes, but included organisational changes, such as staff cuts and training cuts. However, the people at the top of the company appeared to be exempt from this policy. They could order cost cuts without themselves conducting a risk assessment. What I am suggesting here is that the most senior decision-makers of the corporation should be bound by the policy that they have imposed on their subordinates.

Chief executive officers of companies like BP have a strong personal interest in cost cutting. Their remuneration consists (in part) of share options, that is, the right to buy shares at a predetermined price, at some future time. If the share price rises, the executive can exercise the option to buy the shares at the predetermined price and sell them immediately at the higher market price, thereby making a profit. Of course, share prices are strongly influenced by return on investment which, in turn, is influenced by costs. So, all things being equal, a CEO can raise share prices by cutting costs. There is thus a powerful personal incentive for CEOs to drive cost cuts throughout an organisation. What is needed is some equally powerful incentive to ensure that these cost cuts are not at the expense of safety. Perhaps the law should be holding CEOs personally accountable in this respect.

For Eva Rowe, BP's cost cutting was driven by corporate greed. The term "greed" involves a moral evaluation, which is entirely understandable in Eva's case. But we can also provide a more value-neutral analysis by saying that cost cutting at Texas City was driven by a relatively low return on investment. Shareholders require capital to be invested in such a way as to maximise the return on that investment. This is the logic of the capital market. Companies that ignore this logic will fail. From this point of view, the problem was not so much greed, as the failure of senior executives to find more sustainable ways of increasing return on investment, such as closing down certain units or even selling off the site.

Of course, understanding what happened in terms of the logic of the capital market does not preclude moral evaluations.   any observers have characterised the whole economic system as one of institutionalised greed. The problem with such an evaluation is that it offers no way forward. We need to find ways to counteract the costcutting logic by altering reward structures that confront individual decision-makers. One way, suggested here, is to hold CEOs personally accountable for ensuring that cost cuts are not at the expense of safety.

Chapter 9
Reward structures

Organisations often act in ways that seem irrational or contrary to their interests. The failure to prioritise process safety is a case in point. The Texas City accident cost the company billions of dollars and significantly reduced its share price. In short, BP's failure to attend to process safety was economically irrational.

Such irrationalities make a lot more sense, however, when we recognise that organisations themselves don't act — individuals within them do. Behaviour that seems irrational from an organisational point of view may be far more intelligible when seen from the point of view of individual actors. The failure to invest in process safety is much more rational from this point of view. Major accidents are rare and underinvestment can continue for years without giving rise to disaster. On the other hand, managers are judged on their annual performance, especially with respect to profit and loss. Consequently, spending money on process safety is not in their short-term interest. Moreover, business unit leaders tend to think in the short term because they may only be in the position for a couple of years before moving on. Texas City was an extreme case. At the time of the accident, it had had eight managers in the previous five years![1] BP has attempted to address this problem by requiring that business unit leaders stay for longer periods (at least five years in the case of Texas City).[2]

The place of process safety in BP's incentive schemes

Increasing the tenure of business unit leaders is clearly a step in the right direction, but it is not enough. At Texas City, there was a systematic mismatch between the interests of the organisation and the interests of all those working within it. Specifically, the incentive structures that BP had set in place to motivate its personnel did not cover process safety.

BP had what was called variable pay, or a bonus program, that covered all employees. The bonus was based on the overall performance of the refinery; there was no way that individual performance could influence the payment. According to the Chemical Safety and Hazard Investigation Board (CSB), 50% of the bonus was determined by cost leadership (that is, cost cutting), while safety

---

1 *The Report of the BP US Refineries Independent Safety Review Panel* (Baker Report), Washington, US Chemical Safety and Hazard Investigation Board, January 2007, p 34.

2 Baker Report, p 74. The Director of the US Naval Reactor program serves a minimum term of eight years in order to ensure that organisational knowledge is retained. See Columbia Accident Investigation Board, *The Report*, vol 1, Washington, National Aeronautics and Space Administration, August 2003, p 183. On 7 May 2007, *The Times* reported that Shell had recently imposed a job-tenure rule of four to six years on its mid-career and senior executives.

determined only 10%.[3] Furthermore, safety was measured as the number of OSHA-recordable injuries, a measure of personal safety. For workers at the lowest levels of the hierarchy, the bonus amounted to about 6% of base pay, but it increased at higher levels, reaching 40% of base pay in the case of the refinery manager. So while the bonus system was hardly significant for ordinary workers, it was potentially an important motivator for senior managers. Its primary effect was to reward cost cutting, and in so far as it drew attention to safety at all, it was to personal safety rather than process safety. This was a system then that systematically diverted attention from process safety.

Senior managers had an additional system of individually constructed personal performance agreements with their immediate superiors. This system applied to executives above the refinery level and cascaded down to the refinery manager and two levels below. According to the CSB:

> "The contracts consisted of weighted metrics for categories such as financial performance, plant reliability and safety. The largest percentage of the weighting was in financial outcome and cost reduction. The safety metrics included fatalities, days away from work case rate, recordable injuries, and vehicle accidents; process safety metrics were not included. HSE [health safety and environment] metrics typically accounted for less than 20 percent of the total weighting in performance contracts."[4]

Process safety, then, was completely missing from the incentive system at Texas City!

Quite apart from this, what is striking about these bonus schemes is the low weight given to safety in comparison with cost reduction. One wonders in these circumstances whether the bonus system directed much attention to safety at all. The answer appears to be that at Texas City it did. As noted earlier, prior to the accident, the injury rate had improved to such an extent that BP made an additional payment to the whole Texas City workforce.[5] The bonus system did seem to have driven the reported injury rate to very low levels.

This apparently anomalous outcome points to something important about the way bonus systems can work. anagers are influenced not only by economic incentives but also by praise and criticism, and these more subjective influences can be very powerful.[6] oreover, a bonus is not just a monetary incentive; it is a symbolic statement that the recipient has done well. This is especially true if the award of the bonus follows a performance review in which the person concerned

---

3 *Investigation Report: Refinery Explosion and Fire* (CSB Report), Washington, US Chemical Safety and Hazard Investigation Board, March 2007, p 154.

4 Ibid. See also Baker Report, pp 28, 29.

5 It is ironic that this improvement occurred despite three fatalities in that year, two of which were process-related (CSB Report, p 165). The OSHA injury rate excludes fatalities (CSB Report, p 153).

6 Nalbantian, H (ed), *Incentives, cooperation and risk sharing: economic and psychological perspectives on employment contracts*, New York, Rowman & Littlefield, 1987.

is congratulated for performing well according to the specified criteria. In these circumstances, the bonus becomes a psychological reward.

For the most senior people in a corporation, it is hard to see how safety bonuses can provide any significant financial motivation. The financial benefits that they receive go beyond those specified in the performance agreements of most of their subordinates and include payments based on shareholdings and share price movements. Financial returns from these sources can far outweigh other sources of income.[7] In these circumstances, the impact of safety bonuses must be symbolic, affecting reputation and pride, rather than exercising any real financial leverage.

Assuming that incentive schemes do have an effect, either in material or symbolic terms, one of the clearest lessons arising from the Texas City accident is the need to include measures of process safety into remuneration systems. The Baker Panel made the following recommendations:

> "A significant proportion of total compensation of refining line managers and supervisors [should be] contingent on satisfactorily meeting process safety performance indicators and goals . . .

> A significant proportion of the variable pay plan for non-managerial workers . . . [should be] contingent on satisfactorily meeting process safety objectives."[8]

These seem like eminently sensible suggestions.

Some cautionary comments

A warning must be sounded at this point. When deciding on the performance indicators to be included in pay schemes, it is important to recognise that the moment there are consequences attached to performance with respect to an indicator, there is an incentive to manage the indicator itself rather than the phenomenon of which it is supposed to provide an indication.[9]

This is apparent in the case of lost-time injuries. For instance, if people are brought back to work the day after an accident and placed on alternative duties, hey presto, a potential lost-time injury is no longer a lost-time injury. While this can often be justified from an injury management point of view, there is plenty of anecdotal evidence of people being brought back to work purely as a means of

---

7 At Texas City, plant managers were eligible for such benefits, but at least some were only dimly aware of how these benefits were calculated (Lucas, K, deposition, vol 1, 12 December 2005, pp 10, 11). Moreover, they were hardly in a position to influence the amount of the benefit.

8 Baker Report, p 251.

9 See, for instance, Goddard, M, Davies, H, Dawson, D, Mannion, R and McInnes, F, Clinical performance measurement part 2: avoiding pitfalls, *Journal of Royal Society of Medicine* 2002, 95: 549–551.

managing the measure. This was certainly happening at Texas City.[10] any industries have sought to overcome this particular problem by focusing on broader injury categories, such as all injuries requiring any medical treatment. But this, too, is a measure that can be manipulated, and was being manipulated at Texas City.[11] The problem is so severe that a review sponsored by the New South Wales mining industry recently recommended that the industry should no longer pay bonuses on the basis of injury outcome data, such as lost-time or medical-treatment injury rates.[12]

When it comes to process safety, one possibility is to use measures of safety-related *activity*, but such measures are particularly doubtful from the present point of view. The problem is that safety-related activities can vary in terms of their quality as well as their quantity, and it is often possible to increase quantity by sacrificing quality. For instance, if performance is being assessed by the number of audit corrective actions that have been closed out or completed, the result can be improved not only by putting more effort into closing out these actions (the intended outcome), but also by decreasing the quality of the close outs (an unintended outcome). If performance is measured in terms of the number of people who have undergone training, the quantity can be improved by reducing the quality of the training. These are perverse outcomes. They are focused on managing the measure rather than managing safety. In contrast, where a failure rate is being used as a performance indicator, for instance, the rate of leaks or the percentage of pressure relief valves that fail when tested, there is less likelihood of perverse outcomes such as those described above. Of course, a rate of equipment failure can be deliberately understated, but this requires some element of dishonesty, whereas reducing the length of a training program in order to increase the number of people trained involves no such dishonesty. The fact that managers can in good conscience modify safety-related activity in this way is what makes indicators of safety-related activity especially problematic.

This is not to say that performance indicators of safety-related activity should be abandoned; merely that extreme care should be taken before including them in bonus pay systems. Where there is poor performance with respect to such an indicator, the first reaction should be to investigate the reasons. It may be that

---

10　Two such anecdotes are documented in the Telos culture survey that took place at Texas City in late 2004. Here is what one worker wrote: "Auto accident en route to sister site — unavoidable — but we were treated as if we caused the accident even though the other driver was cited. Forced to come to work via taxi when unable to drive and under pain medication which causes drowsiness — bad headache — unable to perform job; management was trying to avoid a lost-time from work. Personal concern was felt, but more concern was given to avoid lost time from work!" (The Telos Group, *BP Texas City site report of findings*, 21 January 2005, p 2).

11　Ibid, pp 3–5. For example, "minor steam burn resulting in first aid visit; management encouraged self-treatment to avoid OSHA recordable injury".

12　*Digging deeper*, Wran Consultancy final report, Orange, NSW Department of Primary Industries, 5 November 2007, p xviii.

there are not sufficient resources to carry out the activity effectively. Penalising poor performance in these circumstances is bound to lead to perverse outcomes.

While it is more meaningful to count failures than activities, this does not mean that counting failures is straightforward. As noted earlier, one of the principal recommendations of the Baker Report was that BP should develop a single summary indicator of process incidents, including fires, explosions, hazardous releases and process-related injuries/fatalities. The report recognised that each of these failure types would need to be carefully defined. So, for instance, it defined "fire" as follows:[13]

- a leak that results in a flame
- a tangible indicator of fire (eg soot on the inside of a distillation column)
- a fire on a scaffold board in a process unit
- a fire in a vehicle parked by an operating unit
- a 120 volt shorted switch
- a ground fault on electrical heat tracing
- a phase-to-ground or phase-to-phase short on electrical power distribution
- a fault in a motor control centre, and
- a fault in electrical switch gear.

The definition also included a list of the types of fire that would *not* be included in this process safety indicator, such as a fire in a rubbish can in an office building. It is clear that even such an apparently discrete and countable event as a fire needs to be carefully defined if it is to be part of an indicator used to drive performance.

These cautionary comments highlight some of the pitfalls when building performance indicators into remuneration systems. This is not an argument against using performance indicators in this way. It is an argument about the need to be constantly alert to unintended effects, that is, attempts to manage the measure rather than managing safety.

Tailoring performance agreements to individuals

The recommendations of the Baker Report distinguish between incentive systems that apply to senior managers and those that apply to employees further down the hierarchy. Clearly, it is senior managers who have the greatest influence on safety, and their incentive structure is therefore critical. In particular, it is critical that their performance agreements focus on matters that are within their power to influence. From this point of view, the problem with many performance indicators is that it is not immediately obvious how the activities of any given manager may influence them. For operations and engineering managers, some of

13   Baker Report, pp 258, 259.

the site-wide performance indicators (such as number of leaks) may be appropriate measures for inclusion in personal performance agreements, precisely because they have direct influence over these matters. The personnel manager, however, may feel that he or she has little or no control over the number of gas leaks, fires or other undesired events occurring at a site. If so, an incentive system that highlights such events has little potential to influence the behaviour of that manager. Putting it another way, measures of this nature are not sufficiently targeted to motivate the performance of such a manager.

On the other hand, it may be that the personnel manager can influence safety by ensuring that positions which are defined as safety-critical are adequately staffed. Such a matter is far too specialised to be included in any incentive scheme applying to other employees, but it may be particularly appropriate in the case of a personnel manager. The point is that senior managers must have performance agreements that identify safety-relevant activities peculiarly within their control and for which they can sensibly be held personally accountable.

oreover, there is no need for these personalised goals to be quantifiable. Some qualitative estimate of the extent to which they have been achieved is all that is necessary in order to be includable in an incentive payment scheme. In one company that I have studied, the CEO writes a series of objectives, some of them safety-related, into the performance agreements of his immediate subordinates. He then evaluates each person in relation to his or her set of objectives on the following scale: meets most; achieves all; exceeds most; exceeds all. Pay is linked to this scale in a predetermined manner.

It may require some ingenuity to devise appropriate safety-related goals for people such as personnel and finance managers. It is vital that this be done, however, because matters within their sphere of influence, such as staff overload and cost cutting, are routinely identified as contributory factors in major accident investigations. Corporate HSE (health, safety and environment) managers can be particularly useful here. Ensuring that appropriate HSE objectives are included in the performance agreements of senior executives is one of the most important functions that a corporate HSE manager can perform. Possible items for inclusion in individual performance agreements of senior managers include:[14]

- actively participate in investigations into all major incidents in area of responsibility
- ensure that health and safety are included in the design of new plant and processes
- attend a health and safety event
- ensure that all immediate subordinates read and discuss a particular book that is relevant to process safety
- include health and safety in all position descriptions

14   Some of these are gleaned from a company that I have studied.

- coach and mentor senior managers on health and safety leadership
- ensure that all positions designated as safety-critical are filled with staff whose competency has been assured
- evaluate the impact of shift arrangements and overtime on levels of fatigue, and
- actively establish that cost cuts are not at the expense of safety.[15]

Once objectives are chosen for the most senior executives, they can be cascaded downwards so that the performance agreements of immediate subordinates encompass matters within their direct control, and so on down the line. In this way, the interests of individuals at all levels can be aligned with those of the organisation.

None of this is to suggest that site-wide or company-wide performance indicators can be dispensed with in the incentive schemes devised for senior executives. If nothing else, these measures serve to remind senior people of overall corporate goals and encourage them to align their behaviour with these goals wherever possible.

A role for regulators

Safety regulators have traditionally seen their role as ensuring that organisations comply with quite prescriptive rules. More recently, some regulators have expanded their role to include auditing safety management systems. There is a further function that they can perform, namely, advising on organisational design. There are numerous aspects of organisational design that have a bearing on safety, one of them being remuneration systems and performance agreements. Such matters have not normally been the focus of regulatory attention. However, given that they can have a profound effect on safety, there is every reason why they should. It would be good to see regulators identifying examples of remuneration systems and performance agreements that emphasise safety and passing these on to companies that have not yet applied themselves to this issue.

Conclusion

Texas City showed a striking incapacity to respond effectively to lessons and messages about the importance of process safety. One of the reasons for this organisational incapacity was that the incentive structures the company had placed before its employees, and in particular its senior managers, systematically diverted attention from process safety. The interests of individuals were not aligned with the interests of the organisation in this respect. If an organisation is serious about process safety, incentive systems for individuals must be designed accordingly.

---

15   See the conclusion to Chapter 8.

Chapter 10
The problem of decentralisation

Both major reports following the Texas City accident argued that BP's decentralised organisational structure undermined process safety.[1] This was not the first time that decentralisation had been identified as a problem for BP. The official report into the Grangemouth events concluded that decentralisation and everything that flowed from it "provided a compelling explanation of the incidents which occurred".[2] Similarly, one of the lessons from Longford concerned the need for head office to take responsibility for the management of major hazards, rather than leaving it to local sites.[3] This chapter will attempt to understand just how decentralisation undermined process safety at Texas City, and will speculate about the ways that organisations might be structured so as to avoid this problem.

I start with an in-principle argument. A decentralised structure means that decisions about how a particular site will operate are made, as far as possible, at the site or local business unit level rather than at head office. This makes sense when it is possible to hold local organisational units accountable for their performance. Individual business units can be held accountable for their business performance, since this is easily measurable. They can even be held accountable for their performance with respect to personal safety, since injury statistics can be compiled and publicised. It is far more difficult to hold them accountable for their process safety performance, since major accidents are rare events. A site may go for many years without a major accident, but it would be foolish to conclude from this that it was managing process safety well.

Furthermore, precisely because individual sites may never have had a major accident, they will naturally become complacent with respect to the management of major hazards. On the other hand, very large corporations are far more likely to have experienced a major accident at some site within the corporate empire. The corporation, in short, has the experience which individual sites may lack. It is also, by virtue of its size, more exposed to the risk of a major accident than any particular site is. In these circumstances, it makes sense for major accident risks

1   *The Report of the BP US Refineries Independent Safety Review Panel* (Baker Report), Washington, US Chemical Safety and Hazard Investigation Board, January 2007, p 94; *Investigation Report: Refinery Explosion and Fire* (CSB Report), Washington, US Chemical Safety and Hazard Investigation Board, March 2007, p 151.

2   Health and Safety Executive, *Major Incident Investigation Report, BP Grangemouth Scotland: 29th May–10th June 2000*, HSE and Scottish Environment Protection Agency, 18 August 2003, p 62.

3   Hopkins, A, *Lessons from Longford: the Esso gas plant explosion*, Sydney, CCH Australia Limited, 2000, pp 34–37.

to be managed in a centralised way by people at head office, rather than leaving it to local managers.

BP demonstrates the point very clearly. It had recently experienced a major accident at Grangemouth. If BP had been managing major accident risks centrally, rather than leaving it to local sites, it is more likely that the lessons from Grangemouth would have been transferred to Texas City.

The Amoco heritage[4]

Process safety at Texas City had not always been managed in a decentralised manner. BP acquired Texas City from Amoco in 1999. Before the acquisition, Amoco had maintained a large corporate HSE (health, safety and environment) organisation that included a process safety group which reported to a senior Amoco vice-president. After the acquisition, BP dismantled the centralised safety structure and responsibility was devolved to the business units. Formal organisational practices were replaced with voluntary groups, such as the process safety "committees of practice". These changes were driven by BP's desire to cut costs, and the unintended outcome was the weakening of process safety voices in the organisation.

Various people within BP saw the Texas City accident as stemming from the days before BP acquired the site. Texas City was often described as a "heritage" site that had not been sufficiently incorporated into the BP system and, in particular, had not aligned itself sufficiently with BP's safety requirements.[5] It would seem, however, that the Amoco heritage gave greater emphasis to process safety than BP's decentralised structure could deliver. Rather than attempting to repudiate the Amoco heritage, BP would have been well advised to embrace this aspect of it.

BP's organisational structure

In order to develop the argument in more detail, let us examine BP's organisational structure at the time of the Texas City accident (see Figure 10.1).

Texas City Refinery was a business unit, which meant that it was a "freestanding commercial organisation", judged primarily in terms of its return on investment.[6] It was answerable for its performance in this respect to a vice-president for United States refining, who in turn answered to a vice-president for refining, worldwide. This person answered to a chief executive for "refining and

---

4   This material is paraphrased from the CSB Report, p 146.

5   Bonse-Geuking, W, deposition, vol 2, 14 February 2007, p 9. A senior company executive expressed this view at a public meeting: "There is one central issue I want to discuss with you — why hadn't BP performance expectations fully taken root at Texas City?" Mogford, J, speech delivered at Orlando, Florida, 24 April 2006.

6   CSB Report, p 147.

marketing", worldwide, who in turn reported to the chief executive of the BP group at the top of the hierarchy. "Refining and marketing" is one of the two main operating divisions of the company, the other being "exploration and production" (not shown explicitly in Figure 10.1). These are respectively the downstream and upstream divisions of the company.[7]

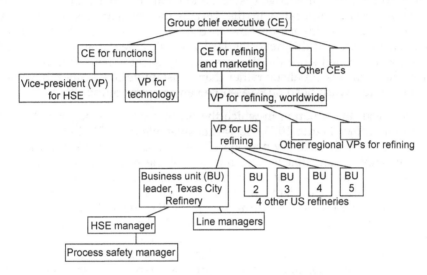

Figure 10.1: BP organisational chart, simplified[8]

Figure 10.1 also depicts a chief executive for functions, with two immediate subordinates, a group vice-president for HSE and a group vice-president for technology.[9] The word "function" refers to an organisational unit within the corporate structure that provides a service to the freestanding business units, but does not itself have any commercial responsibilities. BP's top man, Lord Browne, was very clear on this division. He regularly explained to others that the line was responsible for operations, while the role of the functions was to serve the line, in particular, by developing standards.[10] The functional groups were not responsible for ensuring that these standards were followed — that was the responsibility of the business unit leaders. Putting this another way, standards were determined centrally but responsibility for complying with the standards was decentralised.

7   Oil and gas must first be produced from wells. It is then transported, via pipeline or other means, "downstream" to refineries and then to market.

8   CSB Report, p 148.

9   The position is really vice-president for HSSE (health, safety, security and environment), but to avoid confusion, the second "S" is omitted in this discussion.

10   Hoffman, M, interview for Bonse accountability project, 14 June 2006, p 10; Coleman, G, interview for Bonse accountability project, 26 June 2006. This material is available at www.texascityexplosion.com.

The existence of group vice-presidents for HSE and technology meant that, theoretically, process safety had a champion or champions at a very high level within the corporate structure. In practice, however, these people wielded little influence at site level. For instance, in 2001, the BP group issued a process safety/ integrity management standard that dealt with such things as hazard evaluation, mechanical integrity, protective systems and competent personnel — all matters that were directly implicated in the Texas City accident. The documentation provided detailed guidance on how to comply with the standard. Texas City management reviewed this new standard when it came out and decided that no changes were needed since the site was already in compliance.[11] In retrospect, it is clear that this was ritualistic rather than substantive compliance, but this was not a matter for which head office assumed any responsibility.

Consider, next, the situation in which the process safety manager at Texas City found himself (see Figure 10.1). He was answerable to an HSE manager who in turn answered to the refinery manager. This subordinate position of the process safety manager meant that he was not part of the management team at Texas City and did not, as a matter of course, attend regular management team meetings. He was dependent on his boss, the HSE manager, to represent his concerns at higher levels. As it turns out, the HSE manager had no knowledge of process safety, and no real interest in it. He had never read the federal regulations on process safety management and had no idea what a HAZOP (hazard and operability study) was.[12] He was therefore not in a position to effectively represent the concerns of the process safety manager at a higher level. The result was that the refinery manager was to some extent *shielded* from any concerns that the process safety manager may have had.

Unsatisfactory though this situation was, it was better than it had been. Until a few months before the accident, the process safety manager had not answered to the HSE manager at all, but to a site services manager. In this location, process safety was simply one service in a large amorphous group of services. Even worse, for a period of time, the process safety manager was not directly answerable to the site services manager, being separated from her by yet another layer of management.[13] At this time, then, the process safety manager was three steps down from the site manager.[14]

Commensurate with his lowly status, the process safety manager had a staff of only four so-called "coordinators" working for him. These people had an impossible workload and were quite incapable of carrying it out effectively.[15]

---

11   CSB Report, p 150.

12   Barnes, J, deposition, vol 1, 13 December 2005, pp 10, 11.

13   Hale, R, deposition, 1 June 2006, p 28.

14   Ibid, p 4.

15   CSB Report, p 152.

All of this severely undermined the influence of the process safety manager. His perception was that he did not have the ear of the site manager or even of his own immediate boss. On several occasions, he had urged that process safety should be given a "seat at the management table", but he had been rebuffed.[16] The site manager had been very firm in rejecting this request and, when subsequently asked about this, he said that it was a direction from London.[17]

Finally, there was no way for the process safety manager to raise his concerns at a higher level. In his view, the site manager did not understand the distinction between personal safety and process safety and was not giving sufficient attention to the latter. But there were no lines of reporting between him and the technical people at higher levels in the organisation that would have enabled him to raise this issue. In particular, there was no direct line of reporting to the functional units mentioned earlier. Had there been, he might well have been able to wield more influence.

There is some suggestion in the interview material that one of the reasons the process safety manager had not been given a seat at the table was his personal style; he was seen as merely a technical person and not management material.[18] But, whatever the truth of this suggestion, it is very much a secondary matter. The fact is that BP at Texas City was not committed to ensuring that process safety was represented at the management level by an appropriate person. Had it been, it would have taken steps to find someone with the right combination of technical and managerial skills. Instead, Texas City site managers opted for an organisational structure that systematically disempowered the process safety champion on site. Moreover, they resourced his unit so minimally that it was doomed to be relatively ineffective.

I noted in Chapter 7 that the process safety manager at Texas City had co-authored a paper on the lessons from Grangemouth and that, despite this, the lessons from Grangemouth had had little or no impact at Texas City. The analysis provided here goes a long way towards explaining why.

The preceding discussion suggests two ways in which BP's organisational structure might have been made more sensitive to process safety. First, imagine a direct line between the Texas City business unit leader and a process safety expert under the chief executive for functions (see Figure 10.2). Such a line would make the business unit leader accountable to a functional executive for such things as compliance with process safety standards, and it would require the functional executive to accept some responsibility for ensuring that compliance

---

16   Ralph, W, email, 11 November 2002. He noted that the diversity and inclusion manager reported to the site manager. "[This] leaves no doubt in the mind of the workforce that D&I is a priority and value of the leadership team. In general I would argue that the workforce looks to the organisational chart to assess relative importance."

17   Parus, D, interview for Bonse accountability project, 12 October 2006.

18   Bonse-Gueking, W, deposition, vol 1, 18 October 2006, pp 84, 85; vol 2, 14 February 2007, pp 20, 21; Maclean, C, interview for Bonse accountability project, 26 June 2006.

was occurring, for example, by auditing. The business unit leader under such an arrangement would be subject to dual accountability. He or she would be accountable to the North American refining vice-president for commercial performance, and accountable to process safety experts at head office in London for compliance with process safety standards. A second improvement would be to provide the site process safety manager with a direct reporting line to process safety experts higher in the organisation (see the dotted line in Figure 10.2). These reforms would go some way to centralising responsibility for process safety and remedying some of the problems of the highly decentralised structure that BP had adopted.

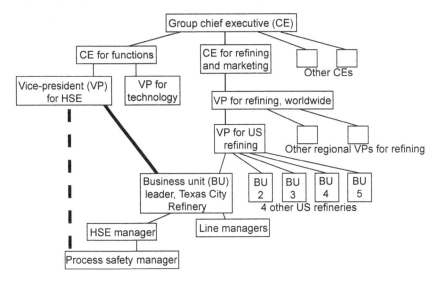

Figure 10.2: Modified organisational chart

Matrix organisation

The preceding discussion identified certain *piecemeal* organisational changes that could be expected to improve process safety. Let us try to think more *systematically* about how this might be achieved. The challenge is to find ways in which a very large corporation can organise itself into separate business units which are responsible for their own business decisions, while at the same time the corporation maintains effective control over standards of operation.

One possible design is the matrix organisation. Interestingly, a BP executive claimed at interview that BP was a matrix organisation. I shall argue later that this was not the case and that, if BP had had a true matrix structure, process safety would have had a far higher priority. So what is a matrix organisation?

Many traditional organisations are divided along functional lines, that is, lines of specialisation. For instance, they may have an engineering department, a design department, a human resources department, and so on. At the same time, the organisation may be engaged in a series of self-contained businesses or projects that have a defined lifetime and/or are carried out at specific locations. Construction projects are an obvious example. Such projects require a project manager who can draw on the specialised resources of the functional groups within the organisation as needed. In these circumstances, the organisation may find it useful to overlay the traditional vertical hierarchy with a horizontal project management structure; hence, the term "matrix". Figure 10.3, taken from the literature on matrix organisations, provides a simple illustration.[19]

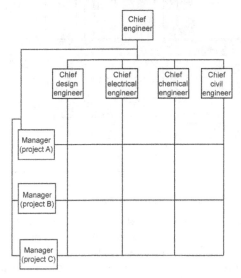

Figure 10.3: A typical matrix organisation

The preceding description presupposes an evolution from a traditional, functionally specialised organisation in the direction of a matrix organisation. It is also possible for organisations to evolve in the opposite direction. Consider a large conglomerate organisation that has grown in much the same way that BP grew, by means of acquiring additional ventures. The conceptual starting point here is a set of independent business units, each with its own functional resources. If the corporation is seeking to standardise operations across business units, indeed, to ensure that the business units conform to corporate standards, it may choose to put the dispersed functional resources of the corporation under centralised control, that is, it may seek to overlay an existing business unit structure with a functional structure. Figure 10.4 is a simplified version of such a

---

19   Ford, R and Randolph, A, Cross-functional structures: a review and integration of matrix
     organisation and project management, *Journal of Management* 1992, 18(2): 267–290, at p 271.

matrix design that was recently adopted by one rapidly growing company in the petroleum industry. It envisages, for instance, a director of operations delivering operational services into all of the business units. Moreover, it envisages a centralised HSE department with expertise in process safety exercising some degree of control over such matters within each business unit.

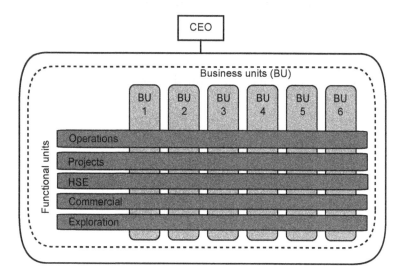

Figure 10.4: A matrix organisation in the petroleum industry

This structure sets up a division of labour: business unit leaders are primarily responsible for the commercial success of their ventures, while functional directors have a responsibility to ensure that their functions are carried out effectively. In principle, in such a structure, the costcutting impulses of business unit leaders will be resisted by functional directors if they conflict with other goals, such as quality and safety. Disagreements about these matters will be debated at the highest level, rather than being resolved in favour of cost cutting at lower levels, as tends to happen in more decentralised organisations. Precisely because these debates are occurring at the highest level, the CEO will be acutely aware of the state of process safety, for example, in contrast to BP's CEO who was not systematically informed on these matters. In these circumstances, process safety will receive an altogether higher priority than it did at Texas City, always assuming that the CEO has the appropriate commitment to safety — an important proviso that will be discussed later.

Notice that the axes in Figures 10.3 and 10.4 are reversed. This reflects the direction in which the organisational evolution has occurred. In Figure 10.3, the starting point is a traditional functionally specialised organisation, on which a project structure has been overlaid, and in Figure 10.4, the starting point is an organisation divided into business units, on which a functional structure has been overlaid. Notice, too, that I have spoken repeatedly of evolution. Research shows that matrix organisations are not installed; they grow from various starting points. Moreover, there is no unique end point in terms of the balance between functional and business units. In Figure 10.3, the project directors report to a chief engineer, which implies that, in the end, functional authority will prevail, while Figure 10.4 implies no clear priority.

Researchers note that one of the advantages of matrix organisations is that they improve the flow of information. In traditional hierarchical organisations, information is filtered and bad news flows upwards with difficulty. In matrix organisations, there are "multiple information channels as a result of the structure".[20] Obviously, a structure such as this would have empowered the process safety manager at Texas City.

A key characteristic of a matrix organisation has yet to be addressed. It is that, in such a structure, employees report to two managers: a functional manager and a business unit or project manager. This is a system that violates the traditional one-boss principle of management, blurs the lines of accountability, and is conducive to conflicts and tensions of various sorts. Critics argue that this is a serious problem, while advocates argue that the tensions are creative, liberating and conducive to adult and responsible behaviour.[21] Either way, it is clear that a matrix organisation introduces ambiguity that needs to be carefully managed.

The concept of a matrix organisation provides one framework for thinking about how BP might seek to redesign itself to give a higher priority to process safety. It renders intelligible the idea of dual accountability suggested in Figure 10.2. The additional lines in that diagram can be seen as a step in the direction of a matrix organisation.

Was BP already a matrix organisation?

I noted earlier that one BP official described BP's organisational structure at the time of the accident as a matrix organisation.[22] The official was BP's most senior executive in North America. He said:

> "We run a matrix organisation and in a classic matrix organisation, the line makes all the operating decisions; and in my case, the functional role is responsible for the fiduciary activity. So, government and public affairs, making sure that our taxes are filed properly, making sure that our financial arrangements with banks are done properly, things like that. That's my role."

---

20 Sy, T and Cote, S, Emotional intelligence: a key ability to succeed in the matrix organization, *The Journal of Management Development* 2004, 23(5/6): 437–455, at p 440.

21 Hunt, J, Is matrix management a recipe for chaos?, *Financial Times*, London, 12 January 1998, p 14.

22 Pillari, R, deposition, 27 June 2006, pp 4, 5, 22, 47. See also Baker Report, p 30.

According to this man, lines of accountability in the organisation were clear and unambiguous. He had never run a refinery and he was in no way responsible for overseeing the refinery at Texas City.

Yet the whole point about a classic matrix organisation as described in the literature is that there are multiple lines of accountability. That is both a strength and a source of ambiguity. What this man was describing was *not* a classic matrix organisation. If anything, it resembled the classic functionally specialised organisation in which he was responsible for ensuring compliance with financial regulations, while others were responsible for operations. Certainly, there was nothing in the structure he described that was likely to promote process safety.

Technical authorities

There is a second organisational strategy for ensuring that commercial considerations do not override issues of technical integrity or process safety, as they did at Texas City. This is the strategy of creating within the organisation a high-level technical authority (or engineering authority or technical regulatory authority, as these authorities are sometimes called). This idea is well discussed in the report of the space shuttle Columbia Accident Investigation Board (CAIB), which is therefore the starting point for this discussion.[23]

According to the CAIB report, NASA (National Aeronautics and Space Administration), the organisation responsible for space flight, was too mission-focused, that is, too focused on launching the shuttle on schedule and insufficiently focused on safety. The organisational structure of NASA effectively disempowered those within the organisation who thought that safety was being compromised. The CAIB turned to the US Navy submarine program for an appropriate model. It noted that the submarine Navy contains a technical authority that is quite independent of normal program managers. Program managers are necessarily sensitive to questions of cost and schedule, while the technical authority is sensitive to safety and technical rigour. Moreover, the technical authority is the "owner" of technical standards and such things as waivers of technical requirements, meaning that it does not merely have an advisory or oversight role, but a decision-making role — it can veto a mission if technical requirements are not satisfactorily met.

Accordingly, the CAIB report recommended the creation of an independent technical engineering authority within NASA. Among its functions were the following:

- develop and maintain all technical standards
- be the sole waiver-granting authority for all technical standards

23   Columbia Accident Investigation Board, *The Report*, vol 1, Washington, National Aeronautics and Space Administration, August 2003, pp 183, 184, 193.

- own the hazard reporting system
- decide what is and is not an anomalous event, and
- independently verify launch readiness.

This is not a complete list of the recommended functions but it gives a flavour of the powerful role that the CAIB envisaged for this authority. In particular, these functions included the power to veto a launch decision if the authority believed that safety requirements had not been met.

Importantly, the CAIB recommended that the authority should be financially independent of the shuttle program:

> "The Technical Engineering Authority should be funded directly from NASA Headquarters, and should have no connection to or responsibility for schedule or program cost."[24]

There is no reason why strong and independent technical authorities, such as the one recommended for NASA by the CAIB, could not be introduced into large decentralised commercial organisations, such as BP. Doing so would be one way of ensuring that commercial considerations did not override process safety.

Finally, it should be noted that the technical authority approach has some overlap with the matrix organisational approach discussed earlier. In a large decentralised organisation consisting of quasi-independent business units, a technical authority is the equivalent of a functional organisation running across the business units. I shall elaborate on this point shortly.

BP reforms

It is interesting to examine some of the reforms that BP has introduced since the Texas City accident in the light of the various models discussed above.[25]

Engineering authorities

One of BP's most important innovations is the creation of a structure of engineering authorities. The following discussion draws on the Baker Report and refers to BP's structure at the time when that report was published in January 2007.[26] To understand this discussion, readers will need to refer to Figure 10.5. This is a severely truncated organisational chart. It includes only those positions necessary to understand the lines of reporting that I shall discuss. For instance, it shows only two positions reporting to the CEO. There are, of course, several others. The engineering authority positions are indicated with heavy borders. It

---

24 Ibid, p 193.

25 There are other changes not discussed here; see Baker Report, p 30.

26 Baker Report, pp 41, 42, Appendix C.

needs to be understood that, in principle, the so-called engineering authority is a single person, although he or she may have support staff.

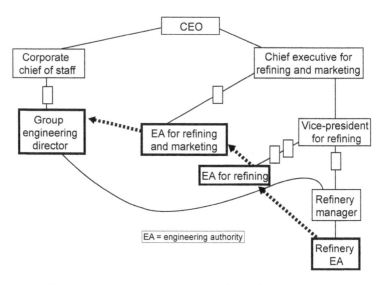

Figure 10.5: BP structure of engineering authorities

At the base of this structure, at the bottom right of the diagram, is a refinery engineering authority. The fact that he reports directly to the refinery manager means that he will exercise more authority than the process safety manager ever did at Texas City. Above him, there are two engineering authority positions, reporting, via various intermediaries, to progressively higher line managers, that is, managers with commercial responsibilities. At the top of this structure is a group engineering director. He is part of the functional arm of the BP group and reports to the CEO via a line that has no commercial responsibilities.

What sets this apart from BP's previous organisational design is the dual reporting lines. According to the Baker Panel, "BP has advised that each site engineering authority will have the authority to raise any issues of concern about what the refinery or refinery plant manager is doing with the refining business engineering authority". In turn, this individual can elevate issues to the next level and, from there, to the group engineering director. This latter individual has the power to override a plant manager. Hence, the curved line in Figure 10.5. In this relatively weak sense, therefore, plant managers have dual lines of accountability: one to their immediate line manager and a weak line to the group engineering director. BP told the Baker Panel that there was frequent communication up and down this line of engineering authorities, that it views this as "engineers talking about engineering", and that this engineering authority line "acts as a check and balance on the authority of the refinery plant manager".

There are various things to be said about this new organisational design. First, although it is described as an engineering authority structure, it is very different from the technical authority found in the Navy and recommended by the CAIB. That authority functioned quite independently from the operational arm of the organisation. In BP, on the other hand, the various levels of the engineering authority are thoroughly integrated into the operational arm and report primarily to line or commercial managers. Only the top position in the engineering authority structure is independent of commercial or line management and able to exercise real authority over line managers. While this adds some independence to the structure, it must also be noted that the top engineering authority position is three steps down from the CEO, by no means an all-powerful position.

In some respects, the BP structure defined above is better described in matrix terms. The engineering authority line cuts across the business units, in much the same way that the functional authority operates in a matrix structure. In particular, plant managers find themselves subject to dual and hence somewhat ambiguous authority. BP's description of the system as one of "checks and balances" recognises and implicitly endorses this ambiguity.

How well all this works, only time will tell. It will depend, in part, on how well BP resources these various engineering authorities. It will depend, too, on the extent to which the group engineering director is willing and able to exercise his authority over plant managers. But they are welcome developments. They reveal an organisation which is struggling to overcome those aspects of decentralisation that contributed to both the Grangemouth and the Texas City accidents.

A senior vice-president responsible for process safety

A second innovation is the creation of a position entitled senior vice-president for safety and operations. What is so important about this position is that it answers directly to the CEO; it is clearly an influential position.[27] Moreover, it is a functional position, that is, independent of commercial or line management. The person who occupies the position explained at interview that his focus was on integrity management, that is, process safety. Prior to this, no one at a very senior level in BP had been specifically responsible for process safety.[28] However, he also stressed that he was not responsible for process safety at site level — that was the responsibility of site managers. This limits the power of the position. In particular, it does not amount to a second line of accountability for plant managers, as proposed in Figure 10.2.

An important part of the new vice-president's job would be auditing, he said. Auditing within the company had previously been done by line management — the same management that was primarily concerned with commercial success. It

---

27   Baker Report, pp 36, 37.

28   Mogford, J, deposition, 29 March 2006, p 53.

was suggested to him at interview that that was like putting the fox in charge of the hens and, while he objected to the analogy, he acknowledged that auditing at Texas City had not prevented the standards there dropping way below BP's expectations.[29] This same man had led the earlier BP accident investigation, which found that:

> "Numerous [internal] audits had been conducted at the site in line with regulatory and corporate requirements, but had generally failed to identify the systemic problems with work practices uncovered by this investigation.
>
> ... these audits focused on documented management systems and processes rather than actual practices, such as following operating procedures. The audits also appear to ignore the history of previous incidents, process excursions, and near misses."[30]

Henceforth, he said, his *external* team would conduct "deep operational audits" at every site, every three years.[31] However, site management would remain responsible for responding to audit findings.

Here, then, is another example of the retreat from the decentralised structure that had allowed things at Texas City to deteriorate so badly. It was described as "increasing independent oversight", while leaving ultimate responsibility for process safety with line management.[32] Nevertheless, one wonders how effective this will be in the long run. Several *external* audits had delivered bad news to Texas City in the past, but nothing much had changed. If the new regime is to make a difference, it will have to find a way to exert real authority over sites so as to ensure that they comply with audit findings.

A possible model is provided by the technical authority responsible for airworthiness in the Australian Air Force.[33] The authority audits Air Force organisations that carry out aircraft maintenance, and it can ultimately withdraw the licence of a maintenance organisation that is not complying with standards. The Australian Air Force talks about black hat and white hat auditing processes — a black hat audit has a policing function, while a white hat audit is more advisory. The audits by the technical authority are definitely black hat, and they are highly influential because of the big stick that the authority carries.

29 Ibid, p 55.

30 *Fatal Accident Investigation Report*, London, BP, 9 December 2005, pp 139, 168.

31 Mogford, op cit, pp 48, 61.

32 Ibid, p 48.

33 *Report of the F111 Deseal/Reseal Board of Inquiry*, Canberra, Royal Australian Air Force, 2 July 2001, pp 8-3, 8-4.

It has been said that BP's new central audit team acts in part as a "coach" and in part as a "cop"; in the preceding terms, it is part white hat and part black hat.[34] It is crucial that the black hat function prevails.

The BP board

One other aspect of BP's organisational structure is relevant at this point, and that is the composition of its board of directors. Both major reports arising out of the Texas City accident concluded that BP's board had not exercised adequate oversight in relation to process safety.[35] In effect, the board had delegated responsibility for process safety down the line but had not ensured that this delegation was being properly carried out. This was yet another aspect of the decentralised approach to process safety that BP had adopted. Symptomatic of this state of affairs is the fact that the head of BP's internal investigation into the Texas City accident was never asked to present his findings to the board. The nearest he got to doing so was a presentation to a board subcommittee. This was the ethics and environmental assurance subcommittee, there being no subcommittee specifically concerned with safety.[36] The Baker Panel recommended that the BP board should take a far greater interest in process safety. The Chemical Safety and Hazard Investigation Board went one step further than this. It recommended that BP appoint "an additional non-executive director of the board of directors with specific professional expertise and experience in refinery operations and process safety".[37] Such an appointment would give process safety a seat at the table at the very top of the corporation.

Management of change

This chapter highlights the importance of managing organisational change from a safety perspective. The need for managing technical change is well understood, but it is clear that major organisational changes, such as mergers, acquisitions and policies of decentralisation, can also have dramatic impacts on safety. Such changes need to be carefully assessed from this point of view. There are numerous industry and regulatory guidelines that spell this out.[38] But BP did follow this guidance. It expected its employees to follow management of change procedures for small-scale changes, such as the location of trailers, but it did not apply the same discipline to itself when it came to staffing cutbacks, major reorganisations and corporate acquisitions. The policy of decentralising

---

34 Broadribb, M, *Three years on from Texas City*, in the proceedings of the US Center for Chemical Process Safety 23rd International Conference, New Orleans, April 2008.

35 Baker Report, p 234; CSB Report, p 189.

36 Mogford, op cit, p 37.

37 CSB Report, pp 214, 215.

38 CSB Report, pp 191, 198, 199.

responsibility for process safety was never subjected to a management of change assessment and the result was that decentralisation undermined process safety at Texas City, just as it had at Grangemouth. BP has now seen the error of its ways and has taken several steps in the direction of recentralising responsibility for process safety.

Chapter 11
Leadership

The Baker Panel wrote eloquently about the failure of BP's top leadership with respect to process safety. Its words are worth quoting:

> "While site leadership is certainly important in establishing a positive process safety culture, the Panel believes that leadership from the top of the company, starting with the Board and going down, is essential. In the Panel's opinion, it is imperative that BP's leadership set the process safety 'tone at the top' of the organisation . . .

> The Panel believes that a primary reason that process safety is not more widely shared as a core value in the US refinery workforce is that BP executive and corporate refining management have not provided effective process safety leadership. Instead, they provided the refining workforce with a plethora of messages concerning many values and these tended to dilute the importance of any corporate vision on safety generally, much less process safety in particular."[1]

In this chapter, I want to discuss two leaders in particular: the CEO of the entire BP group and one of his immediate subordinates, the chief executive for the refining and marketing businesses. Having identified the weaknesses in their leadership with respect to process safety, I shall develop some ideas about the kind of "mindful leadership" that is required to reduce the risk of major accidents.

The CEO

The Baker Panel focused explicitly on the CEO, Lord John Browne of Madingley:

> "The Panel recognises that Browne is a very visible chief executive officer . . . Browne's passion and commitment for climate change is particularly apparent. In hindsight, the Panel believes that if Brown had demonstrated comparable leadership and commitment to process safety, that leadership and commitment would likely have resulted in a higher level of process safety performance in BP's US refineries."[2]

Interestingly, in 2000, a few months after the Grangemouth incidents, Browne wrote a letter about safety in BP's in-house magazine.[3] It contained the following passage:

---

1   *The Report of the BP US Refineries Independent Safety Review Panel* (Baker Report), Washington, US Chemical Safety and Hazard Investigation Board, January 2007, p 60.

2   Ibid, p 67.

3   *Horizon*, BP, October 2000, p 39.

"On a comparison of days-away-from-work case frequencies, we are on par with one of our leading competitors, another is 25% better than us and DuPont is four times better than us.

That shortfall has been emphasised by a series of incidents over the last few months — including serious fires and equipment failures. The result is that our operational integrity has come into question" (emphasis added).

It is clear in this passage that, in Browne's mind, BP's technical integrity problems are linked to the shortfall in personal injury statistics. His failure to distinguish between personal safety and process safety could hardly be more apparent.[4]

One of Browne's colleagues went further than this, claiming that he expressed little interest in safety in general. This colleague had occupied very senior positions. At the time of the accident, he was the BP group vice-president for health, safety, security and environment.[5] He had also held the sought-after position of executive assistant to Browne. He is reported to have said at interview that he had told his CEO many times that he needed to be more involved in safety, but that "John had shown little interest". In association with others, he had proposed several safety initiatives, but Browne had shown "no passion, no curiosity, no interest".[6] The claim here is that the CEO was failing to provide visible leadership on safety in general, not just process safety. If this was how Browne was perceived by a close associate, it is a serious reflection on this aspect of the CEO's leadership.[7]

Furthermore, the view of many observers was that Browne was the type of leader who only wanted to hear good news. According to the man who interviewed various BP managers in preparation for a BBC documentary on Browne, BP had a culture in which many top people knew of problems but few would speak up:

"Only good news flowed upwards", he said. "No one dared say the wrong things or challenge the boss."

And in a revealing comment:

---

4   While Browne did not learn from the Grangemouth accident that process safety needed a separate focus, it seems he did come to understand this after Texas City. Following the release of the Baker Report, he said at a press conference: "If I have one thing which I hope you will all hear today, it is this: BP gets it, and I get it, too . . . I need to take the lead . . . by championing process safety as the foundation of BP's operations" (response to the Baker Report, 16 January 2007).

5   Mogford, J, deposition, 29 March 2006, p 6. This position reported to a deputy CEO for functions who reported directly to Browne (*Investigation Report: Refinery Explosion and Fire*, Washington, US Chemical Safety and Hazard Investigation Board, March 2007, p 148).

6   Coleman, G, interview for Bonse accountability project, 26 June 2006.

7   According to Browne, Coleman subsequently retracted these statements in a letter to lawyers (Browne, J, deposition, 4 April 2008, p 65).

"There were no dissenting voices in the boardroom. He [Browne] was no longer self-aware. Where's his sounding board? Where did he go for feedback?"[8]

The man who was later to succeed Browne expressed the same view:

"We have a leadership style that is too directive and doesn't listen sufficiently well. The top of the organisation doesn't listen sufficiently to what the bottom is saying."[9]

It is clear that, under Browne's leadership, news about the impact that underinvestment at Texas City was having on process safety would not travel very far up the hierarchy.

The chief executive for refining and marketing

Lawyers for those suing BP were not able to interview Browne at length after the accident.[10] They were, however, able to interview his immediate subordinate, the chief executive for the refining and marketing businesses (hereafter, the CE).

Like his boss, the CE looked to injury statistics as the measure of safety. He took comfort from the fact that these were improving.[11] He did not appear to understand that process safety needed to be assessed separately, and he was perceived as being unwilling to spend money on process safety.[12]

The CE was responsible for refining and marketing, worldwide. One of his immediate subordinates had responsibility for the refining side of this business, worldwide. This man was protective of his business and apparently did not welcome close scrutiny by the CE. Indeed, there was a stand-off between these two men. The CE was unable or unwilling to manage his subordinate and he regarded refining as a closed shop — the term used was "fortress refining".[13] This meant, in particular, that he did not inquire closely into how safety was being managed in the refining businesses under his control.

Nevertheless, the CE did make efforts to visit BP's refineries around the world. In 2004, he visited the Texas City site, specifically with safety in mind. There had been a number of incidents and he said:

---

8   Rene Carayol. See article by Steffy, L, *Houston Chronicle*, 14 October 2007.

9   BBC, 1 January 2007.

10   One managed ultimately to conduct an hour-long transatlantic phone call with him (*Houston Chronicle*, 4 April 2008).

11   Manzoni, J, deposition, 8 September 2006, p 28. Unless otherwise indicated, the referenced material is available at www.texascityexplosion.com.

12   Mogford, J, interview for Bonse accountability project, 8 May 2006.

13   Bonse, W, *Management accountability project: supplementary report*, February 2007, p 2.

"I wanted to go and see for myself to make sure that the site was addressing these issues and that [the manager] and his team were on the case."[14]

However, the CE did not get out and about at the plant, and spoke only with the management team. His description of his visit at interview is revealing:

Q: "Are you telling me that no one told you about the problems that existed at the plant during that visit?"

A: "Actually, I left the plant with a sense that programs were being put in place and that there was something called the 1000 day program ... we sat around a table like this and talked about — with the management team — and talked about the issues — and I left the plant with actually a positive impression, not a negative one."

Q: "Someone was giving you a bum scoop, huh?"

A: "Apparently."[15]

He went on:

"I had no indications either that there was any sense of dissent or discontent among the management team."

The CE's visit, then, was an opportunity lost. It was apparently treated by Texas City management as an opportunity to show itself in the best possible light to this senior company executive, rather than an opportunity to talk about the impact that underinvestment was having on the plant. Nor did the CE use this opportunity to check for himself what was really happening at the plant. He was content to accept management assurances. He was a passive recipient of information rather than an active investigator. The problem, it seems, is that the CE had no strategy for getting the best out of his visit. He said later that he "did not know which questions to ask, did not ask the right questions, and was not told [about plant conditions]".[16] This is a critical admission. I shall consider shortly the kinds of questions leaders can most fruitfully ask during site visits.

The CE's general defence under interrogation by hostile lawyers was that he had no idea of the true state of affairs at Texas City. In fact, he said that no one knew just how bad things were:

Q: "Are you telling me that there were not members of management who were quite aware there was a great risk of harm to people at Texas City before this explosion occurred?"

A: "I believe nobody knew the level of risk at Texas City because if they had known, I have absolutely no doubt we would have taken different and

14 Manzoni, op cit, p 4.

15 Ibid, p 42.

16 *The Wall Street Journal — Europe*, 7 May 2007.

substantively different actions. So it's inconceivable to me that we had people who knew that that level of risk existed in Texas City."[17]

The fact is, however, that people did know. As noted in Chapter 7, a major external review of Texas City in 2002 described the failings that it identified as "urgent and far-reaching" and it expressed "serious concerns about the potential for a major site incident".[18]

In 2004, another major review reported that there was an "exceptional degree of fear of catastrophic incidents at Texas City".[19] The CE saw this report after the accident and, at interview, he described its findings as "horrifying" and the situation it described as "completely unacceptable".[20] Then comes a telling interchange:

Q: "Why didn't you find out about these things [before the accident]? . . . you were accountable, right?"

A: "Because in order to find out about them, somebody has to tell me about them."[21]

Nobody, it seems, had told him. This was the CE's ultimate defense. Why had nobody told him? In part, he said, because "the audit process was not functioning properly".[22] But, as we have just seen, the external audit process *was* functioning properly and was coming up with very disturbing findings. What was not functioning properly was the communication system that might have relayed these findings up the hierarchy. So it was that the CE could say:

"There were no audits which were coming to me, for instance, or, indeed, as I understand it, to [my immediate subordinate] which would have indicated the state of that plant."[23]

The CE was three steps above the plant manager in the BP hierarchy and this chain was effectively filtering out the bad news so that none of it reached his level. This was "fortress refining" in action. Bad news was to be kept within the family and not passed up the line. Despite this filtering effect, some bad news may indeed have reached the CE. With regard to the 2002 report, he said: "I may have looked at it but I can't remember it."[24] If he did see it, its "urgent and far-reaching" message was somehow lost on him. But giving him the benefit of the doubt, the best that can be said is that the CE seemed unaware of just how difficult it is for bad news to make its way up an organisational hierarchy. He

---

17 Manzoni, op cit, p 28.

18 *Good practice sharing assessment* (the Veba Report), August 2002, pp 4, 9, 75, 76, 105.

19 The Telos Group, *Executive summary of report of findings*, 21 January 2005, p 2.

20 Manzoni, op cit, p 19.

21 Ibid, p 42.

22 Ibid.

23 Ibid, p 41.

24 Ibid, p 16.

apparently believed that, unless things were brought forcibly to his attention, he could assume that nothing was wrong.

One other matter emerged at interview that casts light on the CE's sense of his leadership responsibilities. At the time of the accident, he had been on holiday with his family in Denver, Colorado. He interrupted his holiday to spend a day at Texas City with the CEO. Subsequently, he wrote an email about this to a colleague. One of the lawyers deposing him forced him to read it aloud for the camera. He read:

> "I arrived in Texas City at 3:00 am along with Lord Browne. And we spent a day there at the cost of a precious day of my leave."

The deposition process aims to identify material that can be put to a jury in the event of a trial, and the lawyer no doubt saw this email as a way of demonstrating to the jury a callous attitude on the part of BP senior executives. My purpose in quoting this email is different. I see it as evidence that the CE felt somewhat remote from the accident and not personally responsible for what had happened.

I once heard a BP business unit leader talk about his responses to the death of a worker who had violated basic safety principles. He said:

> "My first reaction was, 'how could the worker have been so stupid?' Then I realised that he had done what he did in the presence of several work mates, and I thought, 'how could they have let him do this?' Later, I began to wonder about the role of the local manager. Ultimately, I found myself asking, 'why did I let this happen?'"

At the time when the CE sent his email, he was not ready to ask himself this ultimate question.

The preceding paragraphs have described the attitudes and thought processes of one of BP's most senior leaders. It is not a flattering picture. It is a picture of leadership failure. As the internal accountability investigation (to be described later) said about this man:

> "... it is not simply hindsight to suggest that he should have taken more steps to consider and mitigate the risks long before this disaster occurred."[25]

It has only been possible to explore the attitudes of this CE (as we have done here) because the legal system in the United States provides for potential witnesses to be interviewed under oath prior to trial, and because the settlement that BP arrived at with Eva Rowe specified that this material was to be made public. Had Lord Browne, the CEO, been deposed in the same way, it is possible that a similarly unflattering picture would have been revealed. It is not

---

25   Bonse, op cit, p 3.

surprising, therefore, that BP fought tooth and nail to protect Browne from having to undergo this experience. This will be discussed in more detail in Chapter 12.

Mindful leadership

The leadership failures described above challenge us to think carefully about the kind of top leadership behaviour that might have averted the Texas City disaster. I start by outlining the concept of a high-reliability organisation (HRO), because BP aspires to be an HRO. It has consciously adopted the HRO language, and it is seeking to inculcate HRO ways of thinking into its workforce. The idea of an HRO stems from the work of certain academics in the US who studied organisations that appeared to have a remarkably low number of major mishaps. Among the organisations initially studied were aircraft carriers, an air traffic control centre, and a nuclear power station. Researchers found that these organisations were characterised by "collective mindfulness". This meant that, among other things, they were:

- constantly worried about the possibility of failure
- reluctant to draw quick conclusions — always looking for more evidence, and
- sensitive to the experience of frontline operators, encouraging them to speak up.

We can summarise this by saying that mindful organisations exhibit "chronic unease".[26] They embrace the idea that danger may be lurking beneath the surface of normality and that, despite a record of highly reliable functioning, things could go wrong at any moment.

According to HRO theorists, mindful organisations "organise themselves in such a way that they are better able to notice the unexpected in the making and halt its development".[27] As this statement makes clear, collective mindfulness is first and foremost about the style of an organisation, not about the mental state of individuals. It does, however, have implications for the state of mind of individuals at all levels of an organisation. People at the frontline of mindful organisations have an expanded awareness of risk, and leaders of such organisations display a chronic unease about the possibility of things going disastrously wrong. In what follows, I want to develop some ideas about what it means to be a mindful leader.

Mindful leaders lie awake at night worrying about the possibility of a major accident. As one said to me following the Esso gas plant explosion at Longford: "How do we know we are not about to have a 'Longford' at our site? We think

---

26 Reason, J, *Managing the risks of organisational accidents*, Aldershot UK, Ashgate, 1997, p 37.

27 Weick, K and Sutcliffe, K, *Managing the unexpected: assuring high performance in an age of complexity*, San Francisco, Jossey-Bass, 2001, p 3.

we are managing well, but so did Esso." Of course, worry is not always helpful and unease does not automatically generate positive outcomes. We need to think about ways in which this worry can be usefully focused.

Research shows that, prior to every major accident, information was available somewhere in the organisation pointing to the fact that trouble was brewing, but this information failed to make its way upwards to people with the capacity and inclination to take effective action.[28] In short, the bad news that leaders needed to hear did not reach them. The Texas City accident is a classic example of this phenomenon. Mindful leaders develop various ways of identifying this bad news, which I shall discuss below.

Auditing

Senior managers often see auditing as a means of providing themselves with the assurance that things are as they should be. The trouble is that, if leaders are seeking such assurance, that is what they are likely to be given. Auditors may identify what they euphemistically call "improvement opportunities", or "challenges", but if the task is to provide some overall assessment of how well the organisation is being managed, the chances are that the assessment will be positive.

Leaders who want to get beyond these appearances and pinpoint the unrecognised problems that may be lurking beneath the surface need to avoid any suggestion that they are asking for assurances that their system is functioning as intended. Indeed, they need to be suspicious of audit reports that suggest that basically all is well. There are numerous examples of auditors reporting that all is well, only to be followed a few months later by a major accident and an inquiry that reveals that all was not well and that the auditors had failed to identify or highlight some very obvious failings.[29] Where auditors report that all is well, leaders should challenge these assurances and explore with auditors just how they came to these conclusions and how much confidence can be placed in them. They might, for example, pose the following question: "Suppose there were some crucial hazards that remained uncontrolled, or some critical procedures that were not being followed. How likely is it that your audit procedures would uncover these problems?"

One way to overcome this problem is not to ask auditors to provide an overall assessment, but to ask them to identify the most significant safety issues confronting the organisation or site. This gives a radically different purpose to auditing, and it means that auditors who fail to come up with a list of significant concerns have failed in their assignment. No longer is the end point a ranking or a set of numbers; rather, it is an agenda for action, perhaps urgent action. If everyone starts from the assumption that there are likely to be significant

---

28   For instance, Turner, B, *Man-made disasters*, London, Wykeham, 1978.

29   Hopkins, A, *Lessons from Longford: the Esso gas plant explosion*, Sydney, CCH Australia Limited, 2000, Ch 7.

problems and that it is the auditor's job to identify them, then no one need feel undermined when such problems are duly identified.

Auditors who are looking for problems will not approach their task as conventional management system auditors do. They will not set out to sample the organisation. Rather, they will use their expert knowledge to zero in on areas where things might be going wrong, and they will seek advice from others about problem areas. Moreover, they will approach their task with scepticism. They will not merely ask about procedures, they will observe them in action, and they will not necessarily give notice of when they intend to observe what is happening, since that may alter the outcome. Auditors may need to be quite imaginative in the way they follow up their hunches to be certain that they are getting at the truth about what is going on.

Finally, mindful leaders do not simply wait for audit findings to filter up to their level. They commission audits themselves, especially when there are weak signals indicating that all is not well. And they themselves receive and read the reports from these audits. The chief of the Australian Air Force recently demonstrated such an approach.[30] He had received a complaint from a lowly corporal that maintenance activities were not what they should be. He immediately commissioned a special team to visit Air Force bases and report to him on how extensive the problem was. In this way, he was able to bypass the normal Air Force chain of command and inform himself far more directly about what was happening.

From time to time, leaders may become aware of a problem at a site belonging to another company. The problem may have come to light in an accident investigation at that site. Mindful leaders will be concerned that the problem might exist in their own organisation and will appoint an individual or team to find out. For example, the accident sequence at Longford got under way because control room operators were confronted with so many alarms that they could not respond to them effectively. The problem is often described as "alarm overload" or "alarm flooding". When this became known, mindful leaders in organisations around the world worried that control rooms at their sites might suffer a similar alarm overload problem, and they sent investigators to find out. They did not take comfort from the fact that routine safety assurance procedures had not identified such a problem at their site. Rather, they assumed that their routine safety assurance procedures might have missed the problem and that it was therefore necessary to go looking for it specifically. Mindful leaders appoint others to be their eyes and ears and to go out and investigate, bringing to bear the same unease that they feel themselves.

---

30   Author's fieldwork.

Probing for problems personally

The procedures discussed above involve delegation, which necessarily introduces the possibility that information will be distorted or truncated. Mindful leaders know that the best way to find out what is going on is to see for themselves. This means making regular site visits to talk informally with frontline staff about safety issues that they may be facing.[31] But how much time should busy leaders devote to this activity? The report on the Ladbroke Grove rail crash in the United Kingdom addressed this question. What's more, it concerned itself with leaders from top to bottom of an organisation:

> "Companies in the rail industry should be expected to demonstrate that they have, and implement, a system to ensure that senior management spend an adequate time devoted to safety issues, with frontline workers . . . best practice suggests at least one hour per week should be formally scheduled in the diaries of senior executives for this task. Middle ranking managers should have one hour per day devoted to it, and first line managers should spend at least 30 per cent of their time in the field."[32]

Two BP investigations at Texas City after the accident highlighted the importance of leaders at all levels spending time with the workforce. They noted that "the [Texas City management] team [was] not connecting to the workforce in a meaningful way", and "management was generally unaware of local practices".[33] These are devastating criticisms. One report went on to recommend:

> "Leadership team members should schedule time out in the plant (control rooms, craft shops etc) . . . [and they should] listen to the 'voice' of the refinery."[34]

The other report made a similar recommendation:

> "Management team members [must be] more visible on the process units and [must] regularly engage employees in face to face communication."[35]

Interestingly, these investigations did not blame individual managers in this matter. They identified a structural cause of the problem, namely, that there were too few leaders, each with a span of control too great to be able to supervise effectively. As one report said:

---

31 In the management literature, this is described as "management by wandering around". See Peters, T and Waterman, R, *In search of excellence: lessons from America's best run companies*, New York, Harper and Row, 1982. See also, Lavenson, J, How to earn an MBWA degree, *Vital Speeches* 1976, 42: 410–412.

32 Cullen, L, *The Ladbroke Grove Rail Inquiry, Part 2*, Norwich, Her Majesty's Stationery Office, 2001, pp 64, 65.

33 Stanley, J, et al, *Process and Operational Audit Report, BP Texas City*, 15 June 2005, p 5; *Fatal Accident Investigation Report* (FAIR), London, BP, 9 December 2005, p 143.

34 Stanley, op cit, pp 5, 6.

35 FAIR, p 169; see also comments on pp 153, 154.

"The refinery organisation works against senior leadership developing good leadership behaviours due to the extended span of control that exists beneath the senior leaders."[36]

It will be recalled that this extended span of control was a direct consequence of the staff cuts that had been ordered to achieve cost reductions at Texas City.

To return to corporate leadership, leaders who talk to frontline staff must not simply lecture them on the importance of safety — they must listen to them with humility. A few years ago, the offshore oil and gas industry in the UK recognised that its senior executives were spending far too little time offshore, and it mounted a campaign of "boots on for safety". The campaign was accompanied by the image in Figure 11.1.[37] The symbolism was clear — leaders were to listen, not lecture.

Figure 11.1: Image from the "boots on for safety" campaign

Recall that the BP chief executive discussed above did not know what questions to ask. He is not alone in this. Some companies have therefore provided their senior executives with prompt questions. Here is one set of prompt questions used by a major resource company:

- Can you tell me about your job?
- What could go wrong?
- How could you prevent it?
- Who else could be affected?

---

36   Stanley, op cit, p 3.

37   Boots on for safety. Website at http://stepchangeinsafety.net.

- How can the job be done more safely?
- How could you get hurt?
- What kind of injury?

However, there are problems with this particular set of questions. For one thing, it is implicitly focused on personal safety. Moreover, the third question appears to place responsibility for safety exclusively on the frontline worker. This is unlikely to elicit whatever bad news there may be. Consider the following alternative set of questions:

- Can you tell me about your job?
- What are the greatest dangers you face?
- Do you think we have these dangers sufficiently under control?
- Do you think there are any safety issues here that we are not dealing with adequately?
- What would have to be different for you to feel more confident that a disaster at this site is unlikely?
- Are there times when workers feel they need to take short cuts?

This line of questioning does not presuppose that workers are somehow responsible for accidents. It invites them to identify hazards and to consider whether they are adequately controlled. It naturally leads to questions of compliance with procedures. My own experience as an outside observer talking to workers in this way is that they will readily open up in response to these questions, even to the extent of talking about their own non-compliant behaviour.

It is not always easy for a corporate leader to strike up a conversation with a shopfloor worker. The two may be worlds apart in social status, in ways of expressing themselves, and in knowledge and interests. These interactions are potentially awkward for both parties, and leaders may benefit from being coached or from engaging in role play before going out into what may be for them quite a difficult environment.

Leaders who make site visits can also carry out more systematic activities. One is to audit some safety-critical procedure. Auditing should be done with scepticism, imagining ways in which the procedure might be failing, and checking to see whether it is. The famous Piper Alpha disaster in the North Sea in 1988 began with a failure of the permit to work system. The system was regularly audited, but never sceptically, and the systematic flaws in the system were never identified. A leader who identifies flaws in a safety-critical procedure and instructs that they be corrected will have a powerful effect on the culture at that site. Where leaders are unfamiliar with procedures at a site, safety managers can assist by selecting an appropriate procedure and briefing the leader on its purpose and how it should be working.

Another systematic activity for leaders who are visiting a site is to inquire sceptically about the response to some recent accident or near miss at the site. Examining the quality of an accident investigation and the thoroughness with which the site has implemented the relevant lessons provides an important insight into the culture at that site. Far too often, the recommended corrective action arising from an accident investigation is to *talk* to the relevant people to ensure that the problem does not occur again. This is a predictably ineffective strategy. A mindful leader will immediately recognise how such a response leaves the organisation exposed to a repeat accident.

Top leaders who identify problems on site visits should be encouraged to bring them for discussion to whatever management forums they normally attend. One can even imagine company board members engaging in such conversations and bringing their findings to board meetings. Board meetings would never be the same again!

Giving the wrong message

No matter what leaders may say about their priorities, as far as their followers are concerned, their actions speak louder than their words. In particular, followers are very sensitive to what it is that their leaders attend to. The saying, "what interests my boss fascinates me", captures this sensitivity well.

Leaders must therefore be very careful to ensure that the safety messages they think they are giving are not being inadvertently undermined by their own behaviour. For instance, a manager who walks past a situation involving non-compliance with a safety requirement will be perceived as condoning the non-compliance. As the saying goes: "The standard I walk past is the standard I accept." If the admonition to make safety the top priority is to be effective, it must apply to the behaviour of all, from top to bottom of an organisation. Vision statements, safety slogans and lectures are of no use unless they are matched by the behaviour of leaders themselves. If leadership behaviour is not congruent in this way, workers end up believing that it is acceptable to take short cuts — indeed, that management expects them to take short cuts or tolerate certain hazards in order to get the job done.

Mindful leaders are aware of how easy it is to give the wrong messages about safety. The man who took over as the CEO of BP two years after the Texas City accident is a case in point. Shortly after the accident, he gave a speech in which he said: "When a leader visits the workplace, they see the behaviours of their people but they also see, reflected in their people, their own behaviours."[38] Here is a leader who was acutely aware of the subtle messages that his own behaviour might inadvertently convey.

---

38   Hayward, T, BP group managing director, *Working safely: a continuous journey*, in the proceedings of the International Regulators' Forum — Global Offshore Safety, London, 1 April 2005.

Mindful leadership: a summary

To summarise this discussion, mindful leaders do not rely on assurances from subordinates that all is as it should be. They do not assume that the system is functioning as it should in theory. They know, or at least they fear, that there are problems lying in wait to pounce, and they use every means available to probe for these problems and expose them before they can impact detrimentally on the organisation. They focus auditing and other investigative resources on the hunt for such problems. Moreover, they do not merely delegate this task. Mindful leaders regularly visit sites themselves and carry out their own personal audits, looking for indicators that things are not as they should be. They know that if they delegate this investigative function entirely, they can no longer be confident that the bad news will get to them. More generally, mindful leaders welcome bad news. They recognise that it is often difficult to convey bad news upwards and they develop systems to reward the bearers of bad news. Finally, they are aware of how easily their actions can undermine their words about the importance of safety.

Conclusion

Senior executives at BP fell far short of the ideal of the mindful leader that has been outlined above. BP aspired to be an HRO, but its most senior leaders did not recognise that this had implications for their own behaviour. In particular, BP's most senior executives failed to provide appropriate leadership on process safety issues. This is one of the reasons BP seemed unable to implement the principal lesson of previous major accidents, namely, that process safety is distinct from personal safety and needs a specific focus. As we have seen, it was this lack of focus on process safety that led to the Texas City accident.

Chapter 12
Blame

The Texas City accident gave rise to a remarkable amount of blame. Not only were there external legal processes that blamed BP, there were also internal processes in which BP blamed its own employees from top to bottom of the organisation. Some of this internal activity was couched in terms of holding people accountable. The present chapter examines this internal blaming activity and tries to make some sense of the way that accountability and the related idea of responsibility were understood at BP. We shall see that the processes of blame allocation were not only unfair, they were quite bizarre. The chapter concludes with an alternative way of thinking about the accountability of senior executives.

Explanation and blame

Let us begin by distinguishing "explanation" from "blame". An explanation is an attempt to understand what happened and, most importantly, why it happened. In the case of accident analysis, we must ask *why* repeatedly. The answer to every why question provokes another *why*, and these chains of questions and answers are pursued as far as is useful. For instance, if an answer to a why question is "because of a procedural violation", this is followed by another question: "Why did this procedural violation occur?" An answer to this question immediately moves the focus of attention away from the violator and further back along the causal chain. This is a dispassionate activity, aiming for objectivity and seeking to avoid moral or value judgments as far as possible.

It is easy to see how blaming parts company from this approach. Blaming activity looks for violations or failures and, rather than asking why, it passes moral judgment — the person *ought* not to have behaved in this way. The behaviour is labelled wicked, or lazy, or careless, or negligent — all terms which tell us more about the attitudes of the labeller than they do about the behaviour in question. Finally, because the person didn't behave in the way they *ought* to have behaved, they *deserve* punishment.

As can be seen, blaming and explaining are largely incompatible activities.[1] Each is the enemy of the other. Explanation undermines the impulse to blame, because if we understand why someone behaved the way they did there is less tendency to blame them. And blame undermines the impulse to ask why, because blame provides a pseudo-explanation, a root cause of sorts — the person behaved the way they did because they were wicked, or lazy, or careless, or negligent.

---

1   This proposition has been developed at book length by Decker, S, *Just culture*, Aldershot UK, Ashgate, 2007.

This incompatibility means that investigatory bodies must do one or the other, not both.[2] This division of labour can be seen very clearly in the Texas City case. Among the external investigations, the Chemical Safety and Hazard Investigation Board and the Baker Panel aimed to understand, rather than blame, while the court processes, including the Occupational Safety and Health Administration response, were designed to allocate blame.[3] The distinction was mirrored by BP's internal activities. A major internal inquiry that culminated in the *Fatal Accident Investigation Report* aimed to explain, whereas BP's two accountability investigations, described below, aimed to blame, not to understand.

The external blaming processes were all directed at the corporation, not at individuals, and resulted in unprecedented fines. They will be considered further in Chapter 14. What is of central interest here is the internal blame allocation that took place, which was aimed at individuals.

Disciplining the workers

Soon after the accident, BP began disciplinary actions against the workers in charge of the process unit where the accident occurred. According to a prepared statement by the most senior BP executive in North America:

> "... the failure of supervisors to provide appropriate leadership and the failure of hourly workers to follow written procedures are among the root causes of this incident ... We cannot ignore these failures."

Accordingly, six workers were sacked.

The subsequent BP investigation into *management* accountability expressed "full agreement" with these terminations. It described the disciplinary process for the frontline workers as: "... designed to place accountability on those persons directly responsible for the disaster."

These statements are ambiguous. The first describes the failure to follow written procedures as a "root cause". It is hard to imagine any reputable accident analyst describing the failure to follow procedures as a root cause because, as soon we ask why these failures occurred, we are back to more fundamental causes, such as the limitations of the procedures themselves. To describe the failure to follow procedures as a root cause seems like a way of holding the workers responsible for the accident. The second statement makes this explicit; it sees these workers as "directly responsible for the disaster".

---

2   Coronial inquiries, in principle, are set up to explain. They sometimes degenerate into blaming, and the dilemmas and inconsistencies that are sometimes to be found in coroners' reports stem from failing to keep this distinction clear. See honours thesis by Madeleine Rowland, Department of Sociology, Australian National University, 2008.

3   This point is developed in Chapter 14.

We are entitled to ask, then: what were these people being disciplined for? Was it for causing the disaster? If so, this was manifestly unfair. We have seen that the accident was caused by a complex set of factors. If the vent had been replaced years earlier with a flare, the explosion would not have occurred. If the trailers had been sited in accordance with the 350 feet rule, no one would have been killed, and so on. All this can be summarised as a failure by BP to focus on process safety. This, in turn, was due to a series of organisational problems and individual failings at the highest level. To hold workers responsible for the disaster in these circumstances is not only unfair, it downplays the significance of all these other factors.

If pressed to express themselves more carefully, BP spokespeople might possibly deny that the workers were being punished for causing the disaster. Instead, they might say that they were being held accountable for their many failures to do as they were supposed to do. The manager at Texas City who made the decision to terminate these six workers said just that: ". . . the employees were terminated for not fulfilling accountabilities."[4] But if that is the case, why single out these six for discipline when, on all the evidence, the failure to fulfil accountabilities was widespread?

Moreover, instant dismissal is usually only appropriate for malevolent acts, that is, acts done with some kind of malicious intent. In addition, some companies have defined a set of so-called "cardinal rules", for instance, a prohibition on coming to work under the influence of alcohol, and they have warned the workforce that violation of these cardinal rules will result in automatic dismissal. Is this why the Texas City workers were dismissed?

The manager who made the decision to terminate the six workers was asked at interview to explain her reasoning. It seems that she used two distinct guidelines when making her decisions. First, there were indeed offences at Texas City for which workers could be instantly dismissed; she described these as "white card" offences, involving violations of "posted rules". "My recollection is that the six terminations were all on a white card offence", she said.[5] It is not clear on the record just what constituted a white card offence, but what *is* clear is that these offences did *not* result in automatic dismissal. As the manager explained:

> "[A white card offence] is not meant to be the decision-making tool. The decision-making tool is the judgement of the supervisor. The white card offences are a guide. An employee's past behaviours are a guide. There are many things that a supervisor takes into account when taking a disciplinary action."

In short, white card offences were not cardinal offences (as defined above), and to describe behaviour as a white card violation neither explains nor justifies termination. Clearly, something else was going on.

---

4   Lucas, K, deposition, vol 1, 12 December 2005, p 41.

5   Ibid, p 43.

The second guideline used in the disciplinary decision-making was the so-called "just culture policy".[6] This decision-making tool is designed to determine degree of culpability. The original version of this tool was published by Reason in 1997.[7] Various versions are now in use in companies around the world, so it is worth discussing its application in the present case. The Texas City version of the tool is reproduced in Figure 12.1. Readers will need to refer back and forth to the figure to follow the discussion.

The starting point is at the top left of Figure 12.1: "Were the actions as intended?" In the Texas City case, "actions" refer to "violations", and the answer is yes. So the decision-maker moves to the box below and asks: "Were the results as intended?" Assuming that "results" refers to the disaster, the answer is obviously "no".

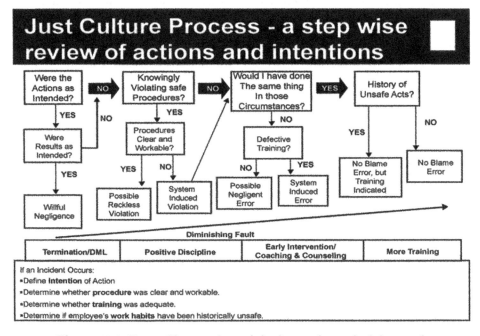

Figure 12.1: Texas City version of the just culture decision tool

We therefore move to the second box in the top row: "[Was the person] knowingly violating safe procedures?" Here, the answer is not so straightforward. Workers knowingly violated procedures, but did they know that these were safety procedures or did they think of them merely as operating procedures? For the sake of argument, let us take the worst possible view, both

---

6   Ibid, p 42.

7   Reason, J, *Managing the risks of organisational accidents*, Aldershot UK, Ashgate, 1997, p 209. This source is never acknowledged in the corporate world.

here and in what follows, and see where it leads. So, assuming that this was a knowing violation of safety procedures, the decision-maker next asks whether the procedures were clear and workable. Assuming that they were, the decision-maker arrives at the conclusion that this was *possibly* a reckless violation. If it was *indeed* a reckless violation, then, according to the next line of the diagram, the behaviour could warrant termination.

But was the behaviour in question reckless? The concept of recklessness has a well-understood meaning at law which was adopted in the original publication by Reason. To be reckless is to engage in behaviour with the knowledge that it involves a likelihood of harm to others.[8] These workers had no idea that their behaviour might result in harm to others — consequently, their behaviour cannot possibly be described as reckless. There is therefore no way that the just culture process should lead to termination in their case, even assuming the worst, as we did above.

But let us go back a step to the question of whether the procedures were "clear and workable". It was argued in Chapter 2 that they were not. In this case, the decision-maker decides that these were system-induced violations and proceeds to ask the question: "Would I have done the same thing in those circumstances?"

There is a problem with the Texas City version of the just culture model at this point. It has to do with the word "I". Given that the "I" is the decision-maker, in this case a senior manager, it is hardly relevant to be asking whether a senior manager would have done as the worker did. Intuitively, it is obvious that we should be asking whether another worker would have behaved in the same way. The problem arises because the Texas City version of the model departs from the Reason original at this point. In the original model, the question is: "[Does the behaviour] pass the substitution test?" The substitution test requires the decision-maker to mentally substitute the individual concerned with someone else *who has the same training and experience* and ask: "In the light of how events unfolded and were perceived by those involved ... is it likely that this new individual would have behaved any differently?"[9] Given that many of the violations were routine, we can say with some confidence that other individuals with similar training and experience would have behaved similarly. This moves the decision-maker to the far right of the diagram, where the behavior is judged to be without blame, but requiring more training.[10]

---

8   Sometimes it is enough to be aware that the consequence is merely possible. Bronitt, S and McSherry, B, *Principles of criminal law* (2nd ed), Sydney, Thomson Lawbook Co, 2005, p 179.

9   Reason, op cit, p 208.

10 The whole just culture model is probably in need of an overhaul so that the substitution test is given greater prominence by asking the substitution question earlier in the decision-making process. If violations are routine, then it probably follows without more analysis that those concerned are relatively blameless. Organisations first need to enforce compliance in a consistent manner, in such a way that violations are not routine. Violations that occur in defiance of an established culture of compliance may then require a more punitive response.

It is perfectly sensible, therefore, using the just culture model, to conclude that the workers concerned were blameless. Even on the worst possible interpretation of their behaviour, there is no way that a decision-maker should arrive at a decision to terminate them.

To summarise, neither the concept of the white card offence nor the just culture model justifies the disciplinary action that was taken against frontline workers. It is clear that, in the minds of the decision-makers, the workers were not simply being held accountable for their violations, they were being held accountable for the disaster. It is hard to escape the conclusion that they were used as scapegoats. In biblical times, the high priest would ceremonially lay the sins of the whole people on the head of a goat, which was then taken into the desert and abandoned.[11] In a similar way, these workers were made to bear a burden of blame that they did not personally deserve.[12]

This discussion should not be interpreted as an argument against holding people accountable for violations. Indeed, the only way that a culture of non-compliance can be turned around is to hold people accountable. The problem is that holding people accountable in circumstances in which there has been an accident almost inevitably involves blaming them for the accident, which is almost certainly unfair. The implication is that, if we want to hold people accountable for their non-compliance, we should do so only in circumstances where there has been no accident. That way we avoid the problem of scapegoating. The consequences for non-compliance can then be measured, starting with warnings and ending up with more punitive outcomes, such as pay reductions. Dismissal may only be appropriate in extreme cases, such as where there is a deliberate refusal to comply.

The Baker Report produced some interesting evidence that workers themselves at Texas City believed that management was not doing enough to discipline fellow workers when they deviated from policies and procedures to do with process safety. Nor was management holding frontline supervisors accountable in this respect.[13] This lack of accountability in the normal course of events is precisely what created the culture of casual compliance, about which Texas City managers complained.

There is one other observation to be made about the just culture model before moving on: it provides an interesting illustration of the distinction between explanation and blame, with which I began this chapter. As we move to the right in Figure 12.1, the behaviour is increasingly *explained* as system-induced or reasonable ("what I would have done"), and the impulse to blame correspondingly diminishes. On the other hand, at the left of Figure 12.1, we

---

11 *Leviticus, The Holy Bible*, King James version, New York, American Bible Society, 1999, Ch 16.

12 They subsequently sued BP for defamation and the cases were settled out of court.

13 *The Report of the BP US Refineries Independent Safety Review Panel* (Baker Report), Washington, US Chemical Safety and Hazard Investigation Board, January 2007, p 94.

have behaviour that is described as "wilful negligence" or "recklessness".[14] These terms imply some level of malicious intent, and the just culture model is content to leave matters there without seeking any further explanation.[15] The premise is that blame is perfectly appropriate if there is no systemic or organisational explanation. We could of course pursue an explanation for such behaviour, using psychology or even psychoanalysis, in which case the impulse to blame would again diminish. But given that, in these cases, an organisation is within its rights to dismiss an offender, there is no need to go down this path.

The management accountability project

Some time after the initial disciplinary process, BP began a second internal accountability investigation, this time in relation to its senior managers. The investigation report recommended that four individuals should be sacked. The language was rather more tactful:

> "The team believes that each of these individuals . . . failed to perform their management accountabilities in significant ways and recommends that BP seek ways to conclude their employment relationships on fair and just terms, in a timely manner."[16]

Given that BP had held frontline workers accountable, it is commendable that the company tried to hold more senior managers accountable as well. But, unfortunately, the management accountability project amounted to a second round of blame allocation, just as problematic as the first. It is worth examining these problems because BP's management system, and that of many other companies, involves delegated accountabilities which must be made to work effectively if accidents like Texas City are to be avoided.

The inquiry was ordered by one of BP's most senior executives, and it was entrusted to another BP executive who was in no way responsible for Texas City. It examined the behaviour of a substantial number of managers — among them, all of the managers who, at the time of the explosion, occupied intermediate

---

14 The "wilful negligence" category is problematic. Negligence means, roughly speaking, carelessness. The idea of *wilful* carelessness doesn't make much sense. This is another instance in which the Texas City model departs from the original. In the original version, the category is "sabotage, malevolent damage, suicide, etc". Interestingly, this description tends to invite explanation rather more than the term "wilful negligence" does.

15 Hart, H and Honoré, T observe that "a deliberate human act is therefore most often a barrier and a goal in tracing back causes" (*Causation in the law*, Oxford, Clarendon, 1985, p 44).

16 *Management accountability project* (Bonse Report), February 2007, p 12. Available at www.texascityexplosion.com.

positions in the line of accountability linking the CEO with the unit in which the accident took place. Specifically, these included:

Chief executive for refining and marketing
Head of refining, worldwide
Head of refining, US
Manager, Texas City Refinery
Assistant manager, Texas City Refinery

The bottom four in this group were recommended for dismissal. The person at the top, the chief executive for refining and marketing (hereafter the CE), escaped judgment. The team was critical of this man but recommended that his "accountability ... should be considered by the appropriate parties within BP". Since the CE reported directly to the CEO, the team was apparently recommending that his case be decided by the CEO.

The questions that immediately come to mind are: why was the CE treated differently, and why did the CEO escape all scrutiny? Is it because their behaviour was more in line with their accountabilities, or is it because of the way the inquiry was set up? The answer is, almost certainly, the latter. The inquiry was initiated by the CE himself and he outranked all of the members on the inquiry team. They therefore found themselves in an awkward position. Should they be considering the accountability of the person who had commissioned the inquiry? The CE made it clear "that the team was free to determine whether he would be included in the scope of its review", and in the end they did include him, although his case was dealt with in a separate report — suggesting just how sensitive the situation was. No doubt, it is for these reasons that the team failed to pass unequivocal judgment on the CE and left it to "the appropriate parties".

As for the CEO, the report says that the team was not given any direction about who it could and couldn't interview. But it is hard to conceive of an inquiry team examining the behaviour of the CEO on the authority of his immediate subordinate. It is highly likely that the team concluded that the CEO was "off limits" as far as the inquiry was concerned. The only way that an inquiry could have effectively examined the accountability of these two men is if it had been set up by the BP board of directors and staffed with at least some people who were not answerable to the CEO. Even then, given the hold that the CEO had over the BP board, it is hard to imagine such an inquiry being frank and fearless when it came to the accountability of the CEO. I shall return to the accountability of the CEO later.

Consider, next, some of the managers whose behaviour was examined but who were found *not* to warrant disciplinary action. One of these had become operations manager at the refinery in January 2005, just a few weeks before the explosion. According to the report, "she simply was not there long enough to have any influence or share any accountability for the disaster". Three other individuals had held senior management roles in the immediate past but had moved to other positions some months prior to the accident. According to the team:

> "... these individuals each had significant management accountability in the past ..., but they had no accountability for the March 23, 2005 disaster ... Their accountability was simply too remote to the incident to warrant any further review."[17]

These statements are very clear: the management accountability project is aiming to identify accountability for the disaster. Managers who performed inadequately were only to be held accountable if that poor performance contributed in some way to the disaster.

However, the report appears to reject this conclusion:

> "None of the management accountability failings identified by the team caused the disaster. Rather, the culture prevalent at Texas City Refinery was the single most direct causal connection."[18]

It goes on:

> "... management accountable for the operations of the Texas City Refinery has to be judged and held accountable for its management shortcomings regardless of whether they had any direct causal impact on the ... disaster."[19]

The report, then, is confused about what managers are being held accountable for. Some people are spared judgment because they are too remote from the incident, while others are to be held accountable for shortcomings, regardless of whether these shortcomings contributed to the incident. Suppose we ask: what precisely was the reason for recommending dismissal? Was it for management shortcomings or was it for contributing to the disaster? The report provides no answer, presumably because the investigative team was never clear in its own mind about what it was doing. This violates one of the basic principles of natural justice: that people know what they are accused of. Moreover, the very process itself was a violation of natural justice. Consider this description:

> "The interviews were not confrontations (at least from the team's perspective) nor were persons questioned in a manner similar to that which occurred in the on-going civil litigation. Rather, the interviews were conversational and the team sought to create a dialogue with each individual to enable the interviewees to discuss what they knew."[20]

---

17 Ibid, p 8.
18 Ibid, p 6.
19 Ibid, p 7.
20 Ibid, p 3.

It is apparent from this description that interviewees were never confronted with any charges and hence were never given an opportunity to marshal a specific defence.

It seems that members of the investigation team went about their business in a naïve way. Without realising it, they systematically denied natural justice to those they interviewed. The fact that intelligent, well-meaning people can flout the principles of natural justice in this way suggests that there is nothing natural about natural justice and that people need to be trained in these principles before they embark on disciplinary inquiries.

Consider, now, the criticisms that the team levelled at the four individuals it regarded as deserving of dismissal.

1. Head of refining, worldwide:
   - promoted a fortress refining mentality by failing to have an effective relationship with the boss
   - never came up with an effective investment strategy to deal with the capital needs at Texas City, for instance, by selling the refinery
   - did not visit the refinery, and
   - failed to respond to warnings contained in the 2002 audit report.

2. Head of refining, US:
   - did not adequately appreciate the process safety implications of the dilapidated state of the Texas City Refinery, and
   - did not effectively manage his subordinate, the Texas City manager.

3. Manager, Texas City:
   - did not adequately understand the distinction between personal safety and process safety
   - did not effectively pass bad news to his superiors
   - was unable to ask for help, and
   - was of questionable effectiveness as a leader.

4. Assistant manager, Texas City:
   - failed to check the competencies of the people in his organisation
   - did not adequately supervise his staff, and
   - had a management style that ultimately led to operational breaches.

I list here, for comparison, the failings of the CE, on whom the team chose not to pass judgment:
   - accepted the fortress refining mentality of his subordinate
   - was unable or unwilling to manage his subordinate
   - failed to debate the appropriate investment strategy for Texas City, and
   - failed to respond to the warnings of the 2002 audit and subsequent audits.

It is clear that the kinds of things of which the CE stood accused were not significantly different from the failings of the other four managers — reinforcing the earlier conclusion that what distinguished him from the others was not his culpability or lack of it but, rather, his seniority.

Many of the criticisms levelled at these individuals are about failures to manage subordinates or to properly inform superiors. These are essentially people management issues. What is striking about these failures is that they are in no way distinctive to this particular line of managers. Many people who work in large organisations have similar experiences. If we applied the substitution test described earlier, this time substituting other senior executives of comparable training and experience, it is hard to imagine their behaviour being very different. The just culture principles suggest that these managers are as undeserving of blame as the workers. The failings identified suggest a lack of managerial competency, not fault. As such, they indicate a need for training, not discipline. These managers need to be taught how to manage people and, in particular, how to hold subordinates accountable. If senior managers are not able to hold people accountable in the normal course of events, organisations inevitably develop silos and pockets of poor performance.

Finally, some of the criticisms concerned a failure to be sufficiently sensitive to process safety issues. Given that the whole corporation, from the CEO down, was deficient in this respect, this again is hardly a matter of individual fault.

To summarise, BP's impulse to hold its managers accountable was understandable, even commendable. But the managers concerned were probably no more blameworthy than many other managers. To see their failings as faults so severe as to warrant dismissal is, again, scapegoating. As in the case of the earlier disciplinary process, if we are seeking to hold managers accountable for their behaviour as managers, it is virtually impossible to do this in a situation in which there has been a major accident. Far better to focus on holding managers accountable in the normal course of events.

On the other hand, if we are willing to entertain the idea of accountability *without fault*, it does indeed make sense to hold the most senior managers accountable for major accidents such as the Texas City disaster. I shall say more about this shortly.

I should record here that BP did not act on the recommendations of the accountability inquiry; no one was dismissed. However, the findings of the inquiry undoubtedly affected the careers of these managers.

Responsibility and accountability

We have yet to focus on the accountability of the CEO, Lord Browne. Leaders create cultures and it is they more than anyone who can change them. There is good reason, therefore, to hold CEOs accountable for the cultures of their organisations and for the consequences of those cultures. The day after the accident, Browne said: ". . . we accept responsibility for . . . this incident."[21] It will be useful to explore the meaning of this statement before looking specifically at the accountability of the CEO.

One of the problems in interpreting Browne's statement is that for BP, and in the corporate world more generally, the concept of responsibility is not quite the same as it is in other contexts. Generally speaking, to be responsible when something goes wrong means to be at fault and/or liable in some way — liable to pay a penalty or perhaps damages. But for BP and its employees, accepting responsibility was not an admission of fault or liability.

Lawyers for those suing BP went to some length to understand what Browne's acceptance of responsibility meant to BP employees:

> Q: "I am just asking you, sir, if you actually know what you mean when you say you take responsibility."
>
> A: "I know what I mean."
>
> Q: "What do you mean?"
>
> A: "I mean that we are responsible for safely operating and the safety of the people that are working on our site."[22]

Here is another interchange:

> Q: "I want to understand what you're saying BP accepts responsibility for."
>
> A: "To determine what happened and take the lessons from what happened and continue to build towards a capability that that would not happen again."[23]

These answers are invoking a subtly different meaning of responsibility, namely, a responsibility is a job, or an aspect of a job. For instance, if I say that marketing is one of my responsibilities, I mean that marketing is part of my job. From this point of view, to accept responsibility for safety on behalf of the company means to recognise that safety is part of BP's job. When Browne was explicitly asked at a press conference following the release of the Baker Panel findings whether he took personal responsibility for the disaster, he said: "I have a responsibility to implement these findings."[24] In so saying, he was acknowledging that he had a job to do, but he was not accepting that he was personally at fault or liable for the accident.

If responsibility means *job*, it is not particularly meaningful to say "we accept responsibility for an *incident*". Indeed, it is quite misleading. Presumably, what Browne intended by his statement was: "We acknowledge that we failed in our job, but through no fault of our own. We tried but we did not succeed. Henceforth, we will try harder."

---

21 See Pillari, R, statement, 17 May 2005.
22 Hoffman, M, deposition, 2 August 2006, p 41.
23 Maclean, C, deposition, 12 September 2006, p 43.
24 Browne, J, response to the Baker Report, 16 January 2007.

It is this subtle difference in the way that the concept of responsibility was understood and used that created such a sense of bewilderment on the part of the lawyers. For them, to accept responsibility without accepting blame was a contradiction in terms.

Of course, BP recognised that some kinds of failures involved fault. But the language used when talking about whether people were at fault was different. The key concept here was accountability. Thus, the disciplinary processes discussed above were seen as inquiries into whether people had acted in accordance with their accountabilities. In short, for BP, and in the corporate world more generally, accountabilities and responsibilities are different.[25]

Nevertheless, the difference is elusive, and in practice there is often no distinction. Consider the following:

> A: "Everything that happens at that plant is the responsibility of the plant manager, who is an employee of the company."
>
> Q: "So are you putting responsibility for this fire and explosion on [the plant manager]?"
>
> A: "No, I am not putting it on anyone. I just said where the accountability was."[26]

The interviewee in this case was one of BP's most senior executives. This is a fascinating exchange — fascinating for the way in which the meaning of responsibility shifts to and fro. Responsibility and accountability are used interchangeably in the first and last line, and in neither case is there any connotation of blame. But when the lawyer asks whether the manager was responsible for the fire, the interviewee takes responsible to mean blameworthy, and the answer is no.

This inquiry into the meaning of the words "responsibility" and "accountability" could be continued indefinitely, but it is not likely to arrive at consistent definitions. Inevitably, meaning must be discerned from context. I shall proceed on that basis.

The accountability of BP's CEO

Consider, now, the accountability/responsibility of the CEO. When Browne said "we accept responsibility", he was referring to BP. Was he also accepting personal responsibility for the accident? At the earlier-mentioned press conference, he had this to say:

> "I think I have a deep and moral responsibility for this company, and in that moral responsibility I always feel that when anything goes wrong, that

---

25   See, for example, Maclean, C, op cit, p 43.

26   Pillari, R, deposition, 27 June 2006, p 11.

> I have let the staff down. I believe that is a moral responsibility of any leader."[27]

This looks at first sight like an acceptance of personal responsibility. But it is not an acceptance that he was personally at fault. There is a certain abstract quality to it: Browne's responsibility is simply the responsibility that falls to any leader when something goes wrong. Browne is accepting responsibility without fault.

Even though BP had its best financial performance ever in 2005, Browne's bonus in that year was 40% less than the previous year, which presumably reflected a board decision to hold him accountable for the accident.[28] This would have been a form of accountability without fault; it is unlikely that the board blamed its CEO personally for the accident. Nor would this kind of accountability have been especially shameful. It amounted to a natural consequence flowing from the formal structure of management accountability.[29]

Accountability to a board is one thing; accountability to the public is quite another. The legal system is the pre-eminent mechanism for holding people publicly accountable. But CEOs or, for that matter, company directors can only be convicted in a court of law if they were significantly at fault, and this is not easy to establish in cases such as the Texas City accident. The result is that CEOs are almost never held publicly to account in these circumstances.

This raises an interesting question. Is there any way that the law can hold CEOs responsible or accountable for major accidents *in the absence of fault*? This is at first sight a bizarre idea. Yet, it can be argued that the law needs to find a way to do just this.[30] Here is a considered statement from the head of a Texas consumer advocacy organisation, Texas Watch:

> "The decisions made by corporate CEOs in board rooms all around the world threaten the safety of communities here in Texas ... [We] all face a greater danger when CEOs are allowed to avoid accountability for the decisions they make.
>
> In Texas, the value of personal responsibility is sacrosanct. Now, I know that Lord Browne isn't from around these parts, but he should still have to answer for his actions. When I was growing up my mother — like parents everywhere — worked hard to instil in me that if I hurt someone — *whether on purpose or by accident* — that I was going to have to face the consequences. The same principle applies to Lord Browne and his band of CEOs.

---

27 Browne, J, response to the Baker Report, 16 January 2007.

28 Baker Report, p 93.

29 Browne resigned from his position as CEO of BP earlier than anticipated, but this had nothing to do with the Texas City accident; it stemmed from an entirely unrelated personal matter for which Browne was both blamed and shamed.

30 This proposition is developed at length in Hopkins, A, *Lessons from Gretley: mindful leadership and the law*, Sydney, CCH Australia Limited, 2007, Ch 7.

When CEOs are allowed to avoid accountability, they will invariably cut corners. When they skimp on workplace safety, then there is a higher likelihood of a catastrophic event . . .

In Texas, no one should be above having to answer for their actions, no one should be allowed to escape accountability, and no one should be able to walk away from dangers they pose without consequence" (emphasis added).[31]

The crucial claim here is that CEOs should be held publicly to account for hurting people, whether it be on purpose or by accident, that is, regardless of whether they are personally at fault. There is widespread community support for this view.[32]

So, how might the law achieve this objective? The legal proceedings that followed the Texas City accident have highlighted one potentially effective way. I refer to the system that exists in the United States of taking depositions from prospective witnesses. Under this system, lawyers for plaintiffs can obtain court orders entitling them to question under oath people at almost every level of a corporation. A deposition can be conducted by a battery of lawyers, can last for many hours, and is filmed. It can be a gruelling and harrowing experience. Plaintiff lawyers who are interrogating senior company executives can be expected to put the behaviour of the witness under a searching spotlight. The process is designed to identify material that can be presented to a jury in proceedings against the company, not proceedings against the executive, but the experience for the executive is one of being on trial for the way in which their actions or inactions contributed to the outcome.

The deposition process holds senior executives to account, in the sense that it requires them to give an account of their behaviour, even though that behaviour may not involve the kind of fault that could result in a criminal conviction. It results in consequences that are just as feared as the consequences that might arise from a formal fault-finding process. We saw in Chapter 11 how the second-in-line to Browne was exposed in this process and ended up being humiliated. Senior executives who are aware of the possibility of such an outcome can be expected to attend to safety rather more conscientiously than they might otherwise do.

Of all the senior BP people in the chain of accountability stretching from Texas City to the CEO, only Browne escaped a full-scale interrogation, despite concerted efforts to have him testify. Lawyers for Browne argued strenuously before various courts that he should not be forced to give evidence. In this they were joined by lawyers from ExxonMobil, among others, who were keen to protect their own CEOs from such an undignified experience. It was not appropriate to subject a CEO to hostile questioning, they argued. To do so would

---

31   Winslow, A, Texas Watch, 18 October 2007.
32   Hopkins, op cit, Ch 7.

make companies "reluctant to shift business operations to Texas" and would undermine "the progress Texas has made in creating an environment hospitable to economic growth".[33] This is a striking admission of how fearful CEOs are of being held to account in this way. It is precisely the kind of accountability that Texas Watch was advocating in the paragraphs quoted above. BP was unable to protect its other senior executives from court-ordered depositions, but Texas law holds that CEOs may be deposed only if it can be shown that they have "unique or superior personal knowledge". BP argued that Browne had no such knowledge. This argument was largely successful: the Texas Supreme Court refused to allow Browne to be deposed in the usual way, but it did authorise lawyers to conduct a one-hour transatlantic phone interview with Browne, a far less stressful experience.

The requirement that CEOs have unique and superior personal knowledge is highly restrictive from the present point of view. The most important questions are not about what Browne knew, but why he *didn't* know about the many problems that budget cutting was causing. The full deposition process had the potential to expose the extent of his lack of attention to these matters. Under BP's system of governance, Browne was ultimately accountable to the board for safety at the Texas City Refinery. A full-scale deposition would have made him accountable to the public as well.

The potential of the deposition process to hold senior executives accountable is only fulfilled if depositions are made public. In the Texas City case, this occurred because of the bargain that Eva Rowe had struck with BP. Were it not for this agreement, there would have been no public accountability at all.

Conclusion

Following the Texas City explosion, BP held accountability inquiries into the behaviour of people from the bottom to almost the top of the organisation. Whether they were being held accountable for the accident or for failing to fulfil their normal obligations was not clear at the time. On reflection, it seems that they were being blamed for the accident. I have argued that this was essentially unfair and that the attempt to hold people accountable after an accident in this way is almost inevitably unfair. Far better to hold people accountable in the normal course of events.

I argued, further, that the process of taking depositions amounted to a form of legal accountability in some cases. Questioning by lawyers revealed how the actions or inactions of some senior people contributed to the accident, and the result was that these people were held accountable *for the accident*, not just for inadequate performance. On the other hand, the process was not one that could

---

33   No 07-0119, Supreme Court of Texas, *In re BP Products North America Inc*, Brief of Amici Curiae, Texas Chemical Council & Ors, p 9.

result in any formal attribution of fault. In this sense, it was accountability without blame.

This is not to belittle its significance. To have one's failings exposed in this way is a shameful experience, and one that is just as likely to bring about behaviour change among senior executives as any formal process of fault-finding. Arguably, accountability processes that mobilise shame are even more effective than those that mobilise blame.[34]

The one person who largely escaped the formal accountability processes following the accident was the CEO. Even though it will seldom be possible to formally attribute blame to CEOs, it is nevertheless beneficial to hold them publicly to account. Eva Rowe had wanted to attend Browne's deposition and show him pictures of her parents' burnt bodies.[35] Had this been possible, it would have brought him face-to-face with his responsibilities in a way that he could never forget. There is considerable potential for the authorities in the US to develop the deposition process into a means of ensuring that CEOs are held publicly accountable. Authorities in other countries need to think imaginatively about other ways of achieving this end.

---

34 This argument is developed in Hopkins, A, *Lessons from Gretley: mindful leadership and the law*, Sydney, CCH Australia Limited, 2007.

35 *Ladies Home Journal*, September 2007, p 164.

Chapter 13
Culture

The culture of BP, and of Texas City in particular, was a culture of blindness to major risk. This was clearly established in earlier chapters. This book could easily have been written as an explicit inquiry into BP's organisational culture.[1] I have chosen not to do so, however, because the concept of culture is often misunderstood. Nevertheless, we do need to think carefully about culture, because BP Refining was actively engaged in what it described as a "culture change program": it was seeking to transform its culture into that of a high-reliability organisation (HRO). Evidently, at Texas City, it had not succeeded. This chapter explains why. The problem is that culture is often used in ways that distract attention from the most important safety issues. Sadly, BP's HRO culture change program fell into this trap. I say "sadly" because the program was in many ways a commendable initiative.

This critique of the HRO program will be approached in stages. First, I shall explore in some detail the meaning of culture in an organisational context. Second, I shall develop the idea of an HRO. This concept was introduced in Chapter 11, but there are certain aspects that need to be emphasised here. Finally, and against this backdrop, I shall pinpoint the limitations of BP's HRO culture change program. There are important lessons here for other companies that have come into contact with HRO theory and seek to transform themselves into HROs.

Description versus explanation

The first problem is that culture can be used either as a descriptive concept or as an explanatory concept. Consider the idea of a culture of risk blindness. To say that a group has a culture of risk blindness is to make a descriptive statement, namely, that people are generally unaware of and insensitive to risk. On the other hand, the statement can also be treated as an explanation for individual cases of behaviour that take no account of risk: they occur because of a general culture of risk blindness.

The term "culture of risk blindness" is useful as a description, since it collects into one category a set of behaviours and attitudes that might not otherwise be linked together. For instance, it groups the behaviour of the operators at Texas City with the behaviour of those who carried out the trailer siting exercise in such a way as to identify a common element. This, in turn, invites us to explain the phenomenon using other concepts, such as the incentive system and inattention to process safety by top leadership.

---

1 Previous accident analyses I have done have been construed as inquiries into cultural causes (Hopkins, A, *Safety, culture and risk*, Sydney, CCH Australia Limited, 2005).

It may also be possible to identify causes of the culture lying outside of the organisation in question. For instance, in contrast to the petrochemical industry, airlines have long been far more sensitive to major hazards than other kinds of hazards. Specifically, they are more focused on the risk of aircraft crashes than on the personal safety of their employees. The reason for this is that public concern about aircraft safety has the potential to put airlines out of business.

While the concept of risk blindness is useful as a description, its value as an explanation is limited. Certainly, it enables us to understand individual behaviour as part of a wider pattern of behaviour, but if we want to change a culture, we need to pursue the causal analysis further to work out why that culture exists. Identifying those causes provides us with potential levers of change. For instance, changing the incentive structure (as discussed in Chapter 9) can be expected to have an impact on a culture that is blind to major risk. From this point of view, the culture of risk blindness is not a root or ultimate cause, but rather an intermediate cause that requires further explanation. As long as we recognise this, there is no problem. But as soon as we treat culture as a sufficient explanation, we run into difficulties. We can see this by recalling the following statement made in the report on management accountability:

> "None of the management accountability failings identified by the team caused the disaster. Rather, the culture prevalent at Texas City Refinery was the single most direct causal connection."[2]

In this statement, the culture at Texas City is seen as the explanation for the behaviour that led to the disaster. Admittedly, the use of the word "direct" leaves open the possibility of indirect causes, but the report does not consider this possibility. Moreover, the above passage appears to discount the possibility that management failings were in any way causal of the accident. In this case, then, the concept of culture is being used in a way that discourages further attempts to understand why the accident occurred.

There is another consequence of stopping the causal analysis once a cultural explanation has been provided. I noted earlier that, once explanation stops, blame takes over. In particular, if we identify a culture of casual compliance as a root cause of the accident, there is a tendency to blame those caught up in that culture. So it was that the report quoted above concluded that frontline workers should be held accountable and punished.

These are some of the problems that can arise when culture is used as an explanation without sufficient attention being paid to the question of where that culture originates. We shall see shortly that the attempt to introduce an HRO culture was undermined by a failure to understand the sources of organisational culture.

---

2 *Management accountability project* (Bonse Report), February 2007, p 6. Available at www.texascityexplosion.com.

Defining culture

The concept of culture has been defined in a multitude of ways: observed behavioural regularities, group norms, espoused values, formal philosophy, rules of the game, climate, embedded skills, habits of thinking, shared meanings, and root metaphors.[3] Some of these definitions focus on values (using the word loosely), but others stress behaviour as the key element of culture. This distinction between values and behaviour corresponds to two common understandings about the meaning of culture: one is that it is about "mindset", the other is that it is about "the way we do things around here". Of course, behaviour is generally informed by values, so there is no actual conflict between these usages; it is simply a question of emphasis.

This distinction is important when we think about the best way to bring about cultural change. Is it better to focus on changing the values or the practices of the group? Organisations frequently seek to change the "hearts and minds" or the "mindsets" of their employees and, in so doing, they are implicitly focusing on values. Often this effort is without much success. The anthropologist Hofstede explains why:

> "Changing collective values of adult people in an intended direction is extremely difficult, if not impossible. Values do change, but not according to someone's master plan. Collective practices, however, depend on organisational characteristics like structures and systems, and can be influenced in more or less predictable ways by changing these."[4]

There is an important consequence of thinking about culture as collective practices. Collective practices can refer to practices of individuals (the way we as individuals do things around here), or they can refer to truly organisational practices that are not reducible to individual practices. Consider the practice of reporting. This is not just a matter of the behaviour of individuals. It depends on the existence of a reporting system; it depends on the organisation providing incentives to report, for instance, recognition and feedback; and so on. These things are vital if a culture of reporting is to be created and maintained, and they are practices of the organisation, not just of individuals.

In order to stress this point, let me briefly describe the reporting system maintained by Australia's air traffic control organisation.[5] The energy that the organisation has put into making this system work is quite remarkable. It does not simply leave it to controllers to decide what to report; it has listed a series of matters that must be reported. In addition, controllers are encouraged to report anything else of concern. Conversations between air traffic controllers and airline pilots are routinely recorded, and the organisation regularly samples these

---

3   Schein, E, *Organisational culture and leadership* (2nd ed), San Francisco, Jossey-Bass, 1992, pp 8, 9.

4   Hofstede, G, *Cultures and organizations*, New York, McGraw-Hill, 1997, p 199.

5   This account is based on my own fieldwork.

recordings to identify reportable issues and to check that controllers are indeed reporting as required. If instances of non-reporting are discovered, these are investigated, and in extreme cases, there may be negative consequences for non-reporters. These practices at the organisational level have created an active reporting culture at the level of individual controllers.

The Baker Panel on BP's culture

It is interesting to note that, when talking about culture, both the Chemical Safety and Hazard Investigation Board (CSB) and the Baker Panel focused on organisational practices. This is nicely illustrated by the Baker Panel's culture survey, which was carried out at BP's five North American refineries.[6] The survey consisted of a questionnaire to which 7,500 employees and contract workers responded. The survey questions were grouped into six categories, as follows:

1. the extent to which the organisation supports and resources the reporting of process safety incidents

2. management commitment to process safety as revealed by its actions, especially in terms of resourcing

3. supervisor behaviour in relation to process safety

4. the adequacy of maintenance

5. worker empowerment, and

6. the adequacy of training.

Clearly, these questions are not about the safety *attitudes* of respondents, nor for the most part about the safety-relevant practices of respondents. They are largely about organisational practices.

A perception survey is not the best way to study organisational practices in any one case, but it is an excellent way of identifying differences between organisations.[7] The Baker Panel survey identified some interesting differences that are worth reporting here because they pinpoint additional sources of the culture of risk blindness that have not so far been recognised.

The first finding of note is that there were indeed significant differences among the five refineries surveyed. Texas City and Toledo stood out as the worst on nearly all measures.[8] At the other extreme, Cherry Point was "the most

---

6   The Baker Panel described its survey as a "process safety culture" survey. I shall try to avoid the term "safety culture" here because it is confusing. See Hopkins, A, *Safety, culture and risk*, Sydney, CCH Australia Limited, 2005, pp 11, 12.

7   For instance, 44% of respondents at Texas City thought that process safety was compromised by financial goals, but it is hard to draw any conclusions from this figure alone.

8   *The Report of the BP US Refineries Independent Safety Review Panel* (Baker Report), Washington, US Chemical Safety and Hazard Investigation Board, January 2007, p 95.

positive".[9] The culture at Whiting was said to be "mixed", while that at Carson was "generally positive".[10] These variations existed despite the fact that all of these refineries were part of the same company and were answerable to the same set of executives as Texas City. We can conclude from this that the factors identified so far are not the only factors influencing refinery cultures. I consider here the "most positive" and the "generally positive" cases, in order to identify some additional factors.[11]

The "most positive" refinery, Cherry Point, had several distinguishing characteristics. First, it was smaller and less complex than most of the others. This meant more interaction and a more family-like atmosphere. Second, it was located in a fairly remote community, which contributed to a sense of inclusiveness. Third, the workforce was largely made up of relatively highly educated, former military personnel from a local military post. Fourth, the leadership had worked to create an egalitarian environment.[12]

With the exception of the last, these factors have a common characteristic: they are beyond the control of plant management and they are beyond the control of the company, in the short to medium term. They are matters that might be taken into account when constructing new refineries, but there is little in the way of guidance about how cultures in existing refineries might be improved.

The refinery with the "generally positive" process safety culture was Carson, in southern California. Industry in this area is highly regulated by various government agencies, and it is this fact that appears to account for the generally positive results in the survey. Here are the words of the Baker Panel:

> "... *on an almost daily basis*, inspectors from any of a number of state or other regulatory agencies might be in the refinery, including the California Division of Occupational Safety and Health, the Los Angeles County Fire Department, Hazardous Materials Division (the Certified Unified Planning Agency for the Carson area), the California Environmental Protection Agency, the California Air Resources Board, and the South Coast Air Quality Management District. Many Carson employees believe, and the Panel does not disagree, that being so closely watched by so many regulators compels Carson to conduct safer operations" (emphasis added).[13]

---

9   Ibid.

10 Ibid, p 104.

11 These were both ARCO refineries before being taken over by BP in 2000 (Baker Report, p 42). One hypothesis is that their good results on the process safety survey reflect this heritage. However, the Baker Report states that "throughout most of the 1990s, while still under ARCO ownership, Carson had a poor safety culture and a deteriorating safety record". Process safety only began to improve in 1998, as result of a commitment by the plant manager at the time (Baker Report, p 100).

12 Baker Report, p 97.

13 Baker Report, p 101. The report lists three other features of the culture at Carson that stand out, but these are descriptive of the culture, not explanatory.

The contrast with Texas City is striking. In the 13 years prior to the accident, the federal Occupational Safety and Health Administration (OSHA) had conducted only one planned process safety inspection at the site, in 1998.[14] There had been other *unplanned* inspections following fatalities at the site. In other words, OSHA was almost entirely reactive in relation to Texas City. There was no proactive program of inspections that might have enhanced the focus on process safety, as occurred at Carson. Similarly, the federal Environmental Protection Agency had never audited compliance with its risk management plan at Texas City.[15] What this suggests is that regulatory activity does indeed have an influence on organisational culture. This is a matter of which governments should take note.

This discussion of the findings of the Baker Panel survey is something of a digression in the present context, although I shall return to the issue of regulation in Chapter 14. The point here is simply that the Baker Panel survey reinforces the idea that the culture of an organisation is best understood as a set of organisational practices that are not necessarily reducible to the practices of individuals — and even less to the mindset of individuals.

What is an HRO?

The preceding discussion has provided us with a useful way of thinking about *culture* as organisational practices. Before discussing BP's attempt to introduce an *HRO* culture, I need to say more about the idea of an HRO. It will be recalled that the essence of the idea is mindfulness. I quote from Weick and Sutcliffe, two academics from whom BP drew inspiration:[16]

> "The key difference between HROs and other organisations in managing the unexpected often occurs in the earliest stages, when the unexpected may give off only weak signals of trouble. The overwhelming tendency is to respond to weak signals with a weak response. Mindfulness preserves the capability to see the significant meaning of weak signals and to give strong responses to weak signals."[17]

As noted in Chapter 11, one of Weick and Sutcliffe's most important observations is that HROs "organise themselves in such a way that they are better able to notice the unexpected in the making and halt its development".[18] This is first and foremost a statement about organisations and organisational practices, not about the mindset of individuals. In other words, if an organisation is to become an

---

14 The *Occupational Safety and Health Act* is administered in California by the state government. In Texas, it is administered by the federal government. See *Investigation Report: Refinery Explosion and Fire* (CSB Report), Washington, US Chemical Safety and Hazard Investigation Board, March 2007, pp 200–209.

15 CSB Report, p 209.

16 *HRO fieldbook*, BP, Ch 2, p 4.

17 Weick, K and Sutcliffe, K, *Managing the unexpected: assuring high performance in an age of complexity*, San Francisco, Jossey-Bass, 2001, p 3.

18 Ibid.

HRO, the first thing its most senior people must do is to put the organisational structures in place that will enable it to see and respond to the unexpected. These structures include reporting systems, auditing systems, training systems, maintenance systems, and so on — all of which have resource implications. Of course, mindful organisations will be populated by mindful individuals, but the mindfulness of most individuals depends on the mindfulness of the organisation.

If we now talk about introducing an HRO *culture*, HRO theory tells us that we are talking about modifying organisational practices, not just the mindsets of individuals. Introducing an HRO culture starts with organisational change. Notice that this is the same concept of culture that was highlighted by the Baker Panel and the CSB.

One of the best examples of the creation of an HRO culture is the US nuclear navy. The emergence of nuclear-powered vessels (particularly submarines) in the 1950s introduced a new potential for catastrophe. This required much greater attention to reliability and safety than previously. The transition to this higher level was driven by a legendary individual, Admiral Rickover, who introduced the training practices, accountability practices, reporting practices, and so on, that characterise HROs. The following passage describes the change:

"The implications of the Rickover culture were far reaching. It forced responsibility down to the operator level. Every reactor operator needed to be aware of what was going on and to make himself responsible for understanding the implications and possible consequences of any action. This intense local awareness by thoroughly trained professionals compensated for ignorance or error at higher levels in the system. It prevented mistakes, some of which might have had catastrophic consequences ... At the time of the Three Mile Island [nuclear power station] accident in 1979, Rickover was in charge of some 152 reactors which had been operating over a period of almost 30 years. Yet no single accident leading to radioactive emission had occurred. In the TMI inquiry, Rickover insisted that this extraordinary reliability was not due to the military context and personnel, rather that it was due to careful selection of highly intelligent and motivated people who were thoroughly trained and then held personally accountable ... [Rickover created a culture in which] communication and recommendations can flow upward from the crewmen to the officers as well as downward. Likewise communication about all kinds of mistakes, operational, technical or administrative, can flow rapidly through the system. Anyone making a mistake can feel free to report it immediately so that the watch officer can really understand what is happening to the system. Rickover believed that the real danger lay in concealing mistakes, for when this happens those in charge become disconnected and disoriented."[19]

---

19   Bierly, P and Spender, J, Culture and high reliability organisations: the case of the nuclear submarine, *Journal of Management* 1995, 21(4): 639–656, at p 651.

I quote this passage at length because, not only does it describe the state of individual mindfulness that is so often associated with HROs, but it also hints at the organisational prerequisites and the enormous organisational commitment that are required to achieve this outcome. The Rickover culture required a redesign of the whole organisation, driven by the commitment of the person at the apex.

BP's HRO culture change program

Against this backdrop, we can now evaluate BP's culture change program — its attempt to convert itself into an HRO, which began in 2000. The thrust of the program can be seen in its HRO "leadership fieldbook", which was distributed to leaders throughout the organisation.[20]

The fieldbook sets out some of the theory of HROs, and it includes an "HRO toolkit" of games, exercises and quizzes aimed at teaching people to think mindfully, that is, to be alert to warnings of danger, to think about "what might bite them", and so on. In short, the HRO program is an educational program, aimed at changing the way people think. It is quite explicit about this: "... cultural change [is about] how people think about themselves, their job and the people they work with."[21] The assumption is that, if people can be educated to think mindfully, BP will be transformed into an HRO.

A memo written by BP's "HRO champion" demonstrates this assumption.[22] The memo was a commentary on an HRO survey that had been conducted among BP employees. It said, among other things:

> "There may be some frustration and cynicism on the part of the front line workers. Front line workers may also have an insufficient understanding of HRO, and may not be effectively engaged (with HRO, operating envelopes, etc), may lack an overview of the intent and purpose of initiatives they work on, and may not really appreciate their own impact and influence.
>
> We thought that a common area for improvement is in developing a better understanding of HRO and engagement of the front line in the behaviours and actions that can have the biggest difference on results."

It is clear from these comments that the whole HRO culture change program was aimed at educating frontline workers to think differently.

The above comments speak of cynicism on the part of the workers. The explanation for this cynicism can be found in the report of a culture survey done at Texas City a few months before the accident. Respondents had learnt the HRO

20 The fieldbook is not only about being an HRO. It includes a major component devoted to teaching leadership skills. I make no comment on that aspect of the fieldbook. The HRO program was introduced only into BP Refining and Pipelines but, for convenience, I refer here just to BP.

21 *HRO fieldbook*, op cit, pp 2–14.

22 Maslin, P, deposition, 13 July 2006, p 65.

language and were willing to talk about weak signals and warning signs. However, in their view the organisation itself was not taking warning signs seriously. Here are some comments from the survey:

"We have warning signs occur every day; like pipe thinning and such."

"Warning signs are everywhere, but the real ones ... [are] the lack of funding and the application of band aids on top of band aids."

"The root cause of the UU4 fire was a lack of sufficient inquiry into weak signals."[23]

Educational programs have their place. But an educational program, by itself, cannot be expected to move the culture of an organisation in an HRO direction. What is required is a different set of organisational practices in relation to training, maintenance, auditing, and so on — all of which lay outside the scope of BP's HRO program.

There was something else that made significant change unlikely. Culture change in organisations starts at the top. The US nuclear navy exemplifies this point. Admiral Rickover's passionate commitment to creating an HRO culture was vital to the outcome. One of the problems with BP's HRO culture change program was that it was not driven from the top. It appears that neither the CEO, nor his immediate subordinate, the chief executive for refining and marketing (the CE), had anything to do with it. Indeed, as we saw in Chapter 11, the behaviour of these leaders was far from mindful: neither was on the look out for "weak signals".

The fieldbook described above was commissioned by an executive who was one step below the CE, and it was developed a little further down the line by an HRO manager. This man described his job as being a "cheerleader for HRO", assisting refinery managers to implement an HRO culture.[24] Here, again, we see the impact of BP's decentralised structure at work. The HRO culture change program was the responsibility of refinery managers, which meant that it had to be funded out of refinery budgets. Given that refineries were under pressure to cut maintenance costs, training costs, staffing costs, and so on, there was really no way that places like Texas City could progress towards HRO status.

Conclusion

There are many ways of thinking about culture, and organisational culture in particular, ranging from individual thought processes, through individual practices, to organisational practices. These things go hand in hand, which has implications for change. Educating frontline workers is unlikely to bring about the desired changes unless organisational practices are also changed in an appropriate way. Given that the HRO culture change program at Texas City was

---

23 The Telos Group, *BP Texas City site report of findings*, 21 January 2005, pp 16, 41.

24 Baker Report, p 41.

largely an educational program, it was never likely to meet with a great deal of success. The prerequisite is organisational change and this requires a commitment at the highest level. This brings us back to individual attitudes again, but this time, the attitudes of the CEO: culture change depends on a change in the thinking of the CEO. It took the Texas City accident to bring about this change. As Lord Browne said at his press conference:

> "If I had to say one thing which I hope you will all hear today it is this: 'BP gets it'. And I get it too. This happened on my watch and, as Chief Executive, I have a responsibility to learn from what has occurred."[25]

Some of the organisational changes described in Chapter 10 result from the fact that the CEO "got it". They are the kinds of changes that can be expected to move BP in the direction of the HRO culture to which it aspires.

---

25   Browne, J, response to the Baker Report, 16 January 2007.

Chapter 14
Regulation

Preceding chapters have made occasional references to the role of regulatory agencies. This chapter aims to bring those comments together in a more systematic way and to propose some strategies for improving regulatory effectiveness.

The regulator in reactive mode

Following the accident at Texas City, the Occupational Safety and Health Administration (OSHA) charged BP with hundreds of violations, such as failure to correct particular equipment deficiencies. The great majority of these were classified as wilful safety violations, or egregious, wilful safety violations. Wilful violations are *a priori* more culpable than non-wilful violations and attract a higher penalty. Labelling them wilful is a way of amplifying the blame.

This point is nicely illustrated by the violation that concerned the use of a vent, rather than a flare. In 1992, this was cited as a *serious* violation, which is less reprehensible than *wilful* in OSHA's scale. The listed penalty was $5,000. The citation was subsequently withdrawn in a negotiation process.[1] In 2005, this same violation was cited as *wilful*, with a penalty of $70,000. The company was now being blamed far more severely for this violation than previously.

The total fine imposed by OSHA was $21m, nearly twice as high as the previous largest fine in OSHA's history. According to a spokesman, the fine "sends a strong message to all employers about the need to protect workers and to make health and safety a core value".[2]

One wonders, however, about the efficacy of such penalties. The previous highest OSHA penalty of something in excess of $10m apparently had no effect on BP, and there is no reason to think that doubling the penalty would make any difference to companies in BP's position. BP made a profit of $19b in 2005.[3] The OSHA fine therefore represented one-tenth of one per cent of the profit in the year of the accident. In any case, the magnitude of the fine is nothing in comparison with the damages that BP has had to pay. There are many circumstances in which fines, including fines for occupational health and safety offences, do have a deterrent or preventive effect, but it is hard to see that this is

1 *Investigation Report: Refinery Explosion and Fire* (CSB Report), Washington, US Chemical Safety and Hazard Investigation Board, March 2007, p 114.

2 OSHA news release, 22 September 2005.

3 Speech by Lord Browne at BP annual general meeting, 20 April 2006.

one of them.[4] I am not arguing here that blaming the company is unfair, as I argued earlier in relation to the blaming of individuals. It is simply that reacting in this way after a major accident is hardly an effective accident prevention strategy.

Moreover, there is something strangely artificial about a multimillion dollar fine that is constructed by adding up hundreds of relatively small penalties. Clearly, what OSHA was trying to do was punish BP for causing the disaster. But OSHA's penalty structure is not designed with that purpose in mind; it is designed as a proactive system in which penalties can be imposed where violations are identified, regardless of whether the violation has resulted in an accident. It is a system designed to prevent accidents, not to punish offenders after accidents have occurred.

The need for proactive scrutiny

In contrast to OSHA's vigorous reaction after the event, the agency failed almost entirely to carry out a proactive program of inspections to ensure that BP was complying with regulatory requirements *before* the accident occurred.[5] Following an explosion at a Phillips plant in Texas in 1989 that killed 23 people, OSHA promulgated a comprehensive process safety management standard.[6] The agency recognised that it needed to carry out inspections against this standard, but in fact it carried out only one planned inspection at Texas City in the 13 years prior to the accident.[7] Despite its best intentions, OSHA simply didn't have the resources to enforce its regulations effectively. This is a matter for which government must ultimately take responsibility.

The Chemical Safety and Hazard Investigation Board (CSB) contrasts the virtual absence of any proactive regulatory activity at Texas City with the situation in the United Kingdom. The UK regulator carries out detailed planned inspections at each of the nine refineries under its jurisdiction *every year*. These inspections range from 80 to 150 days in duration and are carried out by multidisciplinary teams of engineers, instrumentation specialists and human factors specialists.[8]

---

4   Lewis-Beck, M and Alford, J, Can government regulate safety: the coal mine example, *American Political Science Review* 1980, 74: 745–756; Perry, C, Government regulation of coal mine safety: effects of spending under strong and weak law, *American Politics Quarterly* 1982, 40: 303–314; Boden, L, Government regulation of occupational safety: underground coal mine accidents 1973–75, *American Journal of Public Health* 1985, 75(5): 497–501; Gray, W and Scholz, J, Does regulatory enforcement work? A panel analysis of OSHA enforcement, *Law and Society Review* 1993, 27(1): 177–213.

5   CSB Report, p 205.

6   Ibid, p 201.

7   Ibid, p 200.

8   Ibid, p 205.

There is good scientific evidence that intensive regulatory scrutiny is an effective accident reduction strategy. I shall not canvass the evidence here, except to stress again that this is ultimately a resourcing issue: it has been shown that variations over time in accident rates in United States coalmines can be explained almost entirely by the size of the federal budget allocation to coalmine health and safety.[9] It is appropriate to remind readers of what the Baker Panel said on this subject. It found that the process safety culture at Carson refinery was far better than at Texas City, and it suggested that this was because Carson was subjected to far more intensive regulatory scrutiny than Texas City.

It is reasonable to conclude that an absence of regulatory scrutiny at Texas City contributed to the accident or, putting it the other way round, that a more proactive regulatory regime would have reduced the likelihood of the accident.

The need for clear rules

One of the arguments in this book is that, where decisions are dependent on risk assessments, there is an inevitable tendency to downplay the risk in order to enable people to do what they want to do. Where the decision relates to expenditure, risks are likely to be underestimated so as to avoid the expenditure. This tendency to downplay risks was evident both in relation to the trailer siting and the decision not to replace the vent with a flare.

I argued that it is sometimes better to carry out risk assessments remote from the circumstances of particular decisions and to create rules that decision-makers must then comply with. In some cases these might be internal company rules, in some cases they might be contained in industry codes, and in some cases it might be appropriate to formulate them as regulatory requirements.

In particular, where industry best practice is clear and relatively uncontroversial, it may be helpful to formulate this best practice as a regulatory requirement. As an example, the so-called "Eva Bill", which envisages major changes to Texan occupational health and safety law, proposes that temporary buildings should be located at least 1,000 feet from process units.[10] I shall not discuss this particular requirement here. The point is simply that it is much easier for regulators to enforce a clear rule of this nature than a requirement that companies carry out risk assessments. Moreover, given the "licence to operate" mentality of companies like BP, they can be expected to cooperate without much prodding from the regulator — assuming, of course, that the regulator is engaged in a proactive enforcement program. This is not a recommendation for the abandonment of existing legislative frameworks, but it is a suggestion that, in some cases, regulatory objectives may be better achieved by converting risk management requirements into requirements for rule compliance.

---

9   See the references cited in footnote 4.

10   The Bill was defeated in the committee in the Texas legislature, but could still be revived if political circumstances change.

New roles for regulators

Regulatory inspections tend to be focused at the technical level. The analysis in this book highlights organisational issues, which suggests an additional focus for regulators.

In the first place, safety inspectorates could examine the position and powers of company safety specialists. Does the location of safety specialists in the company provide them with sufficient influence? Is there a reporting line for safety specialists at relatively low levels in the company to their counterparts at higher levels? If there are technical authorities in an organisation, do they really have the power to stop a project or an activity that is substandard? Are technical authorities financially independent of business units? Does the corporate centre conduct audits and, if so, how influential are the audit findings?

Regulators would also do well to scrutinise pay incentive schemes and point out to companies how these may be undermining the safety messages they think they are giving. It may also be possible to scrutinise budget priorities for their effects on safety.

Moreover, given that part of the reason for the Texas City accident was that senior executives could claim that they had no knowledge of the real state of affairs at the site, regulators should pay special attention to how well channels of communication are working within a large corporation.

Finally, it may be possible to insist that CEOs apply the same management of change requirements to their own decision-making, particularly with regard to company reorganisations and cost cuts, as is required at lower levels of a company.

These are matters on which inspectorates might simply offer advice or, alternatively, it might be appropriate to issue improvement notices in jurisdictions where this is possible. Otherwise, it may require a change to regulations to facilitate such interventions. For instance, the CSB recommended that the OSHA process safety management standard should be amended to require companies to apply management of change procedures to all organisational changes that may impact on process safety, including:

- major organisational changes, such as mergers, acquisitions or reorganisations
- personnel changes, including changes in staffing levels or staff experience, and
- policy changes, such as budget cutting.

To carry out this expanded role effectively would require inspectors to be trained in matters of organisational design and management. In turn, this would require a substantial increase in the resources provided to regulators. The point is that the root causes of major accidents, such as the Texas City disaster, are to be found at the organisational level in decisions made by senior managers who are remote from the accident. It is desirable that inspectorates have the capacity to identify these root causes and take action to correct them.

Accountability after accidents

The Occupational Safety and Health Administration attempted to hold BP accountable after the accident by "throwing the book" at the company, that is, charging it with every regulatory violation that it could find. In so doing, OSHA was using the system to achieve a purpose that it was not well designed to achieve.

In contrast, BP was also prosecuted by the US Department of Justice under the *Clean Air Act*. The company agreed to plead guilty to a single criminal count. The agreed fine was $50m, which was intended to reflect BP's culpability. This process is a rather more coherent attempt to hold the company accountable for the accident.

However, the agreed penalty amounted to only one-quarter of one per cent of BP's profit for the year. Given that the accident cost BP billions of dollars in other ways, this kind of accountability seems insignificant. A legal challenge to the penalty has been mounted on the grounds that it is too small, and the matter had not been resolved at the time of writing.

Senior officer accountability after accidents

Lawyers for the blast victims were especially critical of the Justice Department settlement for its failure to hold individuals accountable:

> "We believe that unless there is individual corporate management accountability, the horrendous nature of the crime could become significantly downplayed ... We ... established conclusively that management at all levels was repeatedly warned that a tragedy of this nature was inevitable. Those warnings were summarily ignored. It is unconscionable that at least some of those responsible do not go to prison."[11]

A congressional committee investigating the plea also questioned why individuals — particularly the top executives who imposed budget cuts — had escaped liability.[12]

Personal accountability of this kind has the capacity to achieve some degree of closure for victims and relatives. Beyond this, there is plenty of evidence that the fear of personal liability (in particular, the possibility of going to prison) is a powerful motivator, and that CEOs who are aware of this possibility are rather more focused on ensuring that their companies are in compliance with relevant

11   Statement by Brent Coon regarding a decision to postpone the BP plea agreement hearing (see website at www.texascityexplosion.com).

12   *Houston Chronicle*, 12 March 2008.

laws.[13] At the time of writing, it seemed unlikely that anyone would be held personally accountable in this way.

Senior officer accountability without fault

The attempts to depose BP's CEO have highlighted another form of personal accountability. Depositions can hold people accountable — in the sense of requiring them to give an account of their actions and inactions. What is distinctive about this form of accountability is that it is accountability without any presumption of fault. It turns out that CEOs will go to great lengths to avoid being held to account in this way. That very fact suggests that it would be a valuable addition to the regulatory repertoire. Governments should find ways to encourage such accountability.

Once we entertain the idea of accountability without fault, other possibilities emerge. Where a company is prosecuted after an accident and pleads guilty, one of the terms of the agreement might be that the CEO and/or directors meet face-to-face with victims to hear their stories and perhaps to offer an apology. It was just such a face-to-face meeting with the CEO that Eva Rowe had hoped for.

Here is an example that demonstrates the possibilities. It concerns a settlement reached by a regulator that administers laws against misleading commercial conduct. Agents for several insurance companies had been selling worthless life insurance policies to Aboriginal people living in isolated parts of Northern Queensland. The agents had pressured these people in various ways — even threatening that they would be imprisoned if they didn't sign up. Saddest of all, the Aboriginals were falsely told that the policies would pay generous funeral benefits that would help to transport bodies back to the place of origin for burial — a matter of profound importance to them. These matters finally came to court, where senior corporate officers denied knowledge of the practices. Nevertheless:

> "Top management from . . . [one of the companies concerned] were pressed into immediate contact with the victims as part of the process leading up to the settlement. This was an exacting and conscience-searing experience. They had to take four-wheel-drive-vehicles into . . . [the remote outback] to participate in dispute negotiations in which the victims were given an active voice. Living for several days under the same conditions as their victims, [the company's] top brass had to sleep on a mattress on a concrete floor, eat tinned food, and survive at times without electricity."[14]

---

13 See, for instance, Parker, C, *The open corporation*, Cambridge, Cambridge University Press, 2002, pp 92–95.

14 Fisse, B and Braithwaite, J, *Corporations, crime and accountability*, Cambridge, Cambridge University Press, 1993, p 236.

The authors of this passage write eloquently about the consequences of such an experience:

> "Processes of dialogue with those who suffer from acts of irresponsibility are among the most effective ways of bringing home to us as human beings our obligation to take responsibility for our deeds ... Boardrooms ... provide a haven conducive to cosy rationalisations and distorted pictures of actual corporate impacts ... Encountering victims allows the shame of the wrongdoing to be communicated directly to those responsible. The process of encounter also helps to pre-empt or counter efforts by directors or managers to deny the existence of the problem or to neutralise it by means of some self-serving rationalisation. Beyond these salutary effects, encounters with victims provide an opportunity for healing through acceptance of responsibility and putting right the wrong."[15]

Senior officers who are exposed to this kind of experience following workplace deaths will be transformed. Their main focus will no longer be to ensure that they are personally immune from legal liability. They will be motivated to do their best to prevent future accidents.

Managers who have been in charge of a site where someone is killed will tell you that it can be a life-changing experience. They usually know the person killed, sometimes quite well; they are acutely aware of the impact on fellow workers, who may need counseling; and they may have had the unforgettable task of conveying the sad news to the family. What is being proposed above is a legal mechanism for imposing these consequences on senior corporate officers.

The US Department of Justice settlement mentioned above imposed a series of conditions on BP. If such settlements do not include an admission of fault on the part of senior officials, they could at least require senior officers, even board members, to engage in face-to-face dialogue with victims and relatives of victims to hear at first hand about the consequences of their actions and inactions. These meetings would need to be carefully planned to ensure their effectiveness. If well organised, they would not only strengthen the resolve of the most senior people in the company to avoid a recurrence, but they would also provide victims and relatives with a modicum of closure, helping them to move on.[16]

Conclusion

One of the contributing factors to the Texas City accident was inadequate enforcement of health and safety regulations. Had OSHA been enforcing regulations in Texas as intensively as the state administration in California was, the accident would probably have been avoided.

---

15   Ibid.

16   These ideas about accountability without fault are developed at much greater length in Hopkins, A, *Lessons from Gretley: mindful leadership and the law*, Sydney, CCH Australia Limited, 2007.

Governments would do well to require rule compliance, rather than risk assessments, in some cases. This would avoid endless debates about whether certain risk abatement strategies were necessary.

Apart from that, this study has identified aspects of BP's structure and functioning that contributed to the accident. This is where many of the root causes lie. Regulators could profitably focus on these organisational issues, making use of whatever enforcement tools they have at their disposal.

Finally, it was argued that the most effective reactive strategies are likely to be those that target senior officers, either by prosecuting them if they are at fault, or by holding them accountable in some other way if they are not. The idea of holding senior officers accountable without fault is well worth exploring.

Chapter 15
Conclusion

The Texas City accident can be viewed in many ways. For its victims and their lawyers, it was a crime, pure and simple, brought about by corporate greed. The United States Department of Justice also saw it as a case of corporate crime. Various decision-making groups within BP saw it as a failure of individual accountability, from top to bottom of the organisation. The Baker Panel saw it in cultural terms, arguing that BP's process safety culture was defective. In this respect, the Baker Panel was following the lead of the Columbia space shuttle report that attributed that accident to the National Aeronautics and Space Administration's "broken safety culture".[1]

I have chosen to view the Texas City accident another way. The major lesson coming out of the various reports written about the accident is the need for a specific focus on process safety. But this was not a new insight. Time and again, reports about major accidents in the petrochemical industry have drawn this conclusion. What is most striking about the Texas City accident was BP's failure to have learnt this lesson already. It seems that the organisation suffered from a learning disability in this respect.

In this, it was not alone. Accidents often repeat themselves, sometimes within the same organisation. The Columbia space shuttle accident was in some ways a replica of the Challenger space shuttle accident years earlier. The failure of organisations to learn from past accidents is a widespread phenomenon. For this reason, among others, it seemed useful to make the failure to learn a focal point here. This book is therefore, in part, an inquiry into why one organisation, BP, appeared to be incapable of learning from the past.

Figure 15.1 is designed to highlight this failure to learn. The main purpose of this chapter is to summarise the arguments of the book, with reference to Figure 15.1. I shall start with the accident sequence, at the bottom of the diagram. The meaning of the arrows is explained in Appendix 3.

Overfilling the distillation column

The accident began when operators who were starting up a distillation column mistakenly filled it to the point where it overflowed. This is a common mistake and several major accidents around the world have been initiated in this way. The factors that contributed to the mistake on this occasion can be roughly grouped into the following categories:

---

1 Columbia Accident Investigation Board, *The Report*, vol 1, Washington, National Aeronautics and Space Administration, August 2003, p 184.

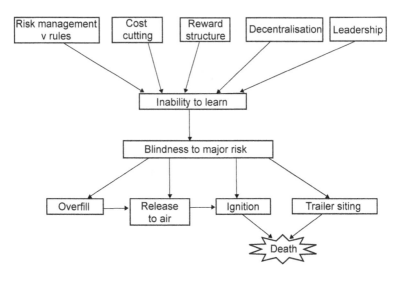

Figure 15.1: A summary of the argument

Supervision and monitoring

- A culture of "casual compliance" in which operators and supervisors treated procedures merely as guidelines that they were free to ignore.
- A failure by BP to audit operator compliance with startup procedures.
- A failure by BP to monitor and react to electronic data on previous startups.

Policies and procedures

- Incomplete and out-of-date procedures.
- A reasonable belief by operators that deviation from procedures was necessary to protect against level fluctuations.
- The lack of a fatigue management policy.

Physical devices and instrumentation

- The lack of any physical cut-out device to prevent overfilling.
- Instruments that only read liquid levels in the bottom section of the column.
- Inadequately maintained instruments.
- A misleading instrument.
- The lack of any display in the control room comparing inflow with outflow.

Communication

- A communication failure between shifts.
- A failure by management to communicate startup instructions.

Training

- The lack of operator training for abnormal conditions.
- The lack of operator awareness of risks.

This is a familiar list. These are the types of things that are routinely identified in major accident investigations. It is a disturbingly long list, but it is not just a catalogue of all of the problems that happened to come to light during the investigation. These are all matters that contributed to the accident. In some cases, we can say quite definitely that, had the matter been otherwise, the accident would not have occurred. For example, had there been a cut-out device that stopped operators overfilling the column, this would certainly have prevented the accident. In other cases, such as fatigue, we cannot have the same level of certainty, but we can say that, had the operators been less fatigued, the accident would have been less likely.

In summary, a host of factors led to the mistake made by the operators. They do not appear in Figure 15.1, for that would make the diagram impossibly complex.[2] Rather, they are summarised in a single box, "overfill". Unpacking this box, as was done in detail in Chapter 2, demonstrated yet again that operator error is best viewed as a starting point for explanation, not as an explanation in its own right.

The remainder of the accident sequence

The next step in the accident sequence in Figure 15.1 is that highly flammable vapour and liquid were released to the atmosphere at the top of a tall vent. This material fell to ground level and accumulated as an explosive vapour cloud. Best practice would have required the material to be ignited or flared at the point of release, so that it could not accumulate and subsequently explode. BP had previously acknowledged this and specified that new plant needed to be equipped with flares. But existing vents could continue to operate. Various opportunities arose over the years to replace the vent with a flare, but these were never taken. Texas City appeared to have no appreciation of the risk that had given rise to the phase-out policy.

The vapour cloud was ignited by a vehicle that had been left idling in the vicinity. The driver of the vehicle apparently had no appreciation of the ignition risk posed by an idling engine, and Texas City had not taken effective steps to minimise this risk.

A number of trailers that were being used as temporary office blocks had been parked much too close to the process unit where the explosion occurred. All of those killed were in or around the trailers. The trailers had been located

---

2 The CSB produced a useful "logic tree" extending over many pages that lays out the causal connections (*Investigation Report: Refinery Explosion and Fire*, Washington, US Chemical Safety and Hazard Investigation Board, March 2007, pp 222ff).

following a risk assessment that failed to take any account of the explosive potential of nearby process units. Indeed, those taking part in the risk assessment had no idea of the risks posed by the process units in the area. The risk assessment, in short, was a travesty.

Blindness to major risk and the inability to learn

All four steps in the accident sequence above have a common thread: a lack of focus on major risk. People at every level seemed unaware of how their actions or inactions might contribute to disaster. At worst, we could describe this as a denial of the significance of major risk; at best, a pervasive blindness to major risk.

The problem was that there was no systematic awareness on site of the need to distinguish between process safety and personal safety. There was no recognition that process safety hazards needed a separate management focus. One of the factors contributing to this lack of awareness was the use of injury statistics as the primary measure of safety. A focus on injury statistics will lead management to focus on the kinds of hazards that are generating injuries most regularly. For instance, if vehicle injuries are prominent in the statistics, this will lead to campaigns to wear seat belts and to drive slowly. The problem is that explosions are rare and do not contribute to injury statistics on an annual basis. An exclusive focus on injury statistics thus leads to complacency with respect to major hazards. This is the trap that Texas City had fallen into, as had so many sites with the potential for a major accident before it. In short, a major lesson arising from the accident was the need to define and prioritise indicators of process safety.

Chapter 7 demonstrated that this lesson was already available and had been brought to the attention of people at Texas City on numerous occasions. Most dramatically, one of BP's own sites, Grangemouth Refinery, had had a series of process incidents that highlighted the need for a specific focus on process safety. But not even this lesson within the company had impacted on Texas City. This was so striking a failure that it seemed reasonable to describe BP as suffering from a learning disability. This is a somewhat metaphorical term, of little explanatory value in itself, but it does call attention to the fact that there is something here so remarkable as to require further explanation. Much of the book is devoted to this task.

Of risk and rules

One thing that contributed to the failure to learn is that process safety was treated as a matter of risk management and not rule compliance. This was government policy, as described in Chapter 14. It meant, for example, that there was no specific regulatory rule that required Texas City to replace the vent with a flare.

Risk management was also the philosophy by which the company operated. Unfortunately, it was a philosophy that impeded organisational learning at Texas City. An obvious way to learn from previous accidents is to devise rules that must henceforth be complied with. For instance, knowing that previous accidents had been caused by overfilling distillation columns, Texas City might have developed a rule that all distillation columns were to be fitted with automatic cut-out devices. That way, learning from previous accidents would be embedded in the organisation. But the philosophy at Texas City, and in the industry more generally, is that individual cases can be individually risk managed. Unless great care is taken, these individual risk management processes can result in substandard decisions and, in particular, decisions that take no account of lessons from elsewhere.

The trailer siting policy provides a particularly clear example of the problem. Corporate risk engineers had calculated that 350 feet from potential explosion sources was a safe distance. But the policy did not impose this as a rule. It allowed trailers to be closer, provided a local risk management exercise was undertaken. This exercise turned out to be entirely ineffective; it was a ritualistic exercise to justify a decision that had already been taken on grounds of convenience.

The other problem was that BP's budget priorities undermined the risk management approach. There were three budget categories. First priority was expenditure required to bring a site into compliance with regulatory requirements; second priority was expenditure required to sustain production; and third priority was spending to take advantage of new commercial opportunities. Spending to reduce risk does not feature in this scheme, so any project the purpose of which was to reduce risk would be unlikely to receive funding. A decision to replace the vent with a flare could never be justified on the basis of these priorities. In short, the risk management philosophy, coupled with a set of budget priorities that undermined risk management, made it very difficult for Texas City to implement the lessons of previous accidents which required far greater expenditure on process safety.

A final related point. Senior BP executives were accustomed to saying that they would spend whatever was necessary to ensure safety. This commitment assumes that safety is an "either/or" matter, things are either safe or not safe. If we are dealing with clear safety rules, then the commitment translates into spending whatever it takes to ensure compliance. However, the risk management framework assumes a continuum of risk, with no clear dividing line between safe and unsafe. In this context, senior management commitment to spend whatever it takes to ensure safety has no clear meaning. It is presumably sincere, but nevertheless empty, rhetoric.

Cost cutting

Cost cutting was perhaps the most obvious cause of the failure to learn at Texas City. The site had been paralysed by years of cost cutting and did not have the capacity to respond to lessons from elsewhere, or even to explicit warnings provided by audit reports. There were too few staff at management levels to supervise frontline employees, there was not enough money to fix or install safety devices and instrumentation, and so on. Crucially, there was not enough money to equip Texas City with simulators that would have enabled operators to be trained in how to handle abnormal process conditions.

Texas City had been hit with numerous requests over the years to cut its costs, the most infamous being the 25% budget cut that London headquarters ordered when BP acquired Texas City in 1999. This was ordered without regard to the safety implications; people at the top of the corporation simply assumed, without applying their minds, that safety would not be compromised.

Although Texas City was returning a large profit, it was a particular target of cost cutting because it was a very large site, tying up a very large amount of capital — and the return on this capital was not in line with BP's expectations. The rational thing to do in these circumstances was to sell the site. However, BP senior executives did not do this, preferring to hold on to what in some ways was its flagship, and to demand further cost cuts. The result was that Texas City was in a seriously run-down condition by the time the accident happened.

One of the contributors to BP's costcutting mentality was comparative industry "benchmarking" data that suggested that Texas City was not as cheap to run as some other refineries. These data do not take account of the particular circumstances of a refinery that may require additional expenditure. Nevertheless, the benchmark data fuelled demands from senior executives for cost cuts.

Incentives

When an organisation behaves in ways that seem irrational, as Texas City did by downplaying process safety, we need to examine the incentive system that operates for senior staff. At Texas City, these incentives were predominantly focused on financial outcomes, including how effectively costcutting targets were met. One component of bonus pay was determined by safety performance, as measured by injury data. There were no process safety indicators included in these incentive schemes. As a result, the incentive system drove managers to focus on personal safety, rather than process safety.

In short, the lessons from previous major accidents were pointing in one direction, while the incentive system was pointing in another. It is not surprising that senior managers seemed so unreceptive to those lessons.

Decentralisation

BP had a decentralised organisational structure. This meant that business units such as Texas City were relatively autonomous in how they operated, subject to an overriding requirement that they provide a satisfactory return on investment. How they dealt with process safety was largely a matter for site management to determine. This meant that the relentless pressure from senior executives to cut costs was not counterbalanced by any comparable pressure from higher corporate levels to ensure that the site operated safely.

There were process safety specialists at higher levels in the company, but they had no authority and no influence over sites like Texas City. There was also a process safety specialist at Texas City, but he was low in the pecking order and wielded little influence with site management. If he had had a reporting line to higher technical authorities in the company, and if those authorities had had authority at site level, things might have been different — but that would have violated the organisational model that BP had adopted.

Chapter 10 discussed some alternative models that would give process safety more clout. One of these was an independent technical authority answerable to the CEO. This is a model used in some high-reliability organisations (HROs). It amounts to an organisation within an organisation, which has no responsibility other than for safety, and which has the capacity to override commercial managers anywhere in the broader organisation who are judged not to be managing major risk effectively. BP has since adopted some aspects of this model with its structure of engineering authorities, that is, people at various levels of the organisation who are the custodians of technical standards. But these people are not independent of line management and so do not have the freedom from commercial pressures required by the technical authority model. Nevertheless, the new structure means that decision-makers at site level have dual accountabilities: they are accountable to commercial line managers, but they are also accountable to a line of engineering authorities running up the organisational hierarchy. In this respect, the new structure has some of the characteristics of a matrix organisation. Unfortunately, the top of this line of engineering authorities is three steps down from the CEO, which significantly limits the influence of the position.

BP has taken other steps since the accident to recentralise responsibility for process safety to some degree. These will not be discussed here. They all involve a recognition that the decentralised structure that the company had adopted was detrimental to process safety and had effectively prevented sites like Texas City from implementing lessons from elsewhere, even lessons from BP's own site at Grangemouth.

Leadership

The final factor that hampered the ability of the organisation to learn was the attitude of its most senior leaders. The CEO, in particular, did not understand the difference between process safety and personal safety. Moreover, in the opinion of one of his closest associates, he showed little interest in safety in general. He did not encourage bad news about safety to move up the hierarchy; indeed, he was perceived as wanting to hear only good news. He was not a leader who was mindful of the possibility of major accidents. His mind was very much focused on commercial matters and climate change issues, and he was content to leave safety to others. This created an environment in which it was very difficult for the organisation to implement the lessons from earlier accidents.

Chapter 11 developed some ideas about mindful leadership that will not be repeated here. Suffice it to say that mindful leaders do not assume that no news is good news, as BP's leaders did. They know that there is always some bad news and they strive to uncover it — by setting up organisational mechanisms to do so and by doing their best to find out for themselves.

Regulation

There is one other factor that contributed to the accident, but it does not appear in Figure 15.1 — inadequate regulatory oversight. Chapter 14 argued that a more proactive regulatory regime of inspections would have forced management at Texas City to focus more effectively on process safety, and that this would have made the accident less likely. There is an important lesson here for governments. It is not enough to enact regulatory requirements, such as the process safety management standard. These requirements need to be enforced. This, in turn, means that governments must resource their inspectorates sufficiently well to be able to carry out this function.

Although inadequate regulation contributed to the accident, it has not been included in Figure 15.1 because it cannot be argued that it contributed to the failure to learn. All other elements at the top of the diagram actively impeded BP's capacity to learn; regulatory failure is a contributory factor of a different kind.[3]

---

3   Figure 15.1 could have been drawn in such a way as to include this factor. While it did not contribute to BP's learning disability, it did contribute to the persistence of a culture that was blind to major risk. But the diagram is not intended to be a complete causal explanation; rather, it is a summary of the book's organising argument about failure to learn. For this reason, regulatory failure is omitted.

Blame and accountability

The Texas City accident and its aftermath provide us with a rich case study in the processes of blame and accountability. To start with, blaming must be contrasted with explaining: the former involves moral judgment, the latter is an attempt to understand why. These activities are largely incompatible; each undermines the other. In particular, blaming workers for the mistakes that contributed to an accident tends to undermine attempts to understand why they made those mistakes. Moreover, blaming workers is fundamentally unfair — as became clear with BP's accountability exercise in relation to its frontline workers. These workers were simply the last link in a long chain of events.

Of course, employees need to be held accountable for their behaviour, but it is far better to do so in the absence of any accident, that is, in the normal course of events. Consequences can then be proportionate to the degree of culpability of the violation itself, rather than to the severity of the harm. The obstacle to holding people accountable in the normal course of events is that the violations must be detected, and this requires supervisory resources that were not available at BP. Companies that are serious about accountability need to recognise these resource implications.

What is fascinating about the BP case is that senior managers were blamed as well. The process involved a refreshing acknowledgment that, if frontline workers were being held accountable, it was appropriate to hold senior managers similarly accountable. However, the process was similarly unfair. The accountability report stated that these managers were not being held accountable for the accident, but for various quite nebulous failures in the way they carried out their management responsibilities. These failures were hardly sackable offences; they suggested the need for training, not discipline. Again, it would have been much better to pursue these matters in the normal course of events, rather than in the shadow of an accident.

This raises the question of whether it is ever appropriate to hold individuals accountable for a major accident. I focus here on the question of legally imposed accountability. Consider, first, the possibility that people are negligent with respect to the consequences of their behaviour, that is, that they should have been aware of the possible consequences but weren't.[4] In these circumstances, punishment can be justified as having some deterrent value. In cases such as Texas City, however, it is difficult to establish that people should have been aware that their actions or inactions might lead to disaster.

Alternatively, courts could find ways to impose accountability *without* fault. This would be most appropriate in the case of very senior officers, such as CEOs. In these circumstances, it would not be appropriate to punish, but other

---

4   An even worse case is that they were reckless, that is, they were aware of the possible or likely consequences but didn't care.

consequences might be devised to confront CEOs with the end results of their actions or inactions. For instance, where companies are found guilty, settlements might include a requirement that the CEO meet with families of the victims to hear their pain and perhaps to apologise for their own part in what happened. This could be expected to have a significant impact on their future behaviour.[5]

Culture

Culture is another concept that is woven inextricably into this story, in part because the official analyses draw on this idea, but also because BP itself had embarked on a culture change program. BP was trying to develop the culture of an HRO. I argued that it had failed, at least at Texas City, because of BP's failure to understand the essence of organisational cultures and where they come from. Culture change must be driven from the top of the organisation. In this case, the change was not driven from the top and was left to site managers to implement. Moreover, the program was understood as an educational program designed to get individuals to be more risk aware. But organisational cultures are best understood as a set of organisational practices, rather than the attitudes of the individuals who make up the organisation. Culture change therefore requires, first and foremost, changes in the way that the organisation does things, not changes in the way that people think. Educational campaigns are of limited value from this point of view.[6]

The special nature of the BP Texas City case

Accident analyses are generally richest in information about the events on the day of the accident and the role of those at the front line. Good analyses go beyond this and identify organisational failures that contributed to the accident. Few analyses identify the contribution of the most senior people in the corporation. In the present case, the trail goes to the very top of the corporation, the CEO. This is largely possible because of the quality of the investigations that have been done, and because of the system in the US of deposing potential witnesses in civil cases, which generates priceless insights into the way senior people think and act. All of the information acquired for the purposes of civil litigation was made public as a result of an agreement with BP negotiated by a single plaintiff, Eva Rowe. She was insistent that the material be made public so that others might learn from it. I trust this book goes some small way towards fulfilling her hopes.

---

5   These issues are explored in much greater detail in Hopkins, A, *Lessons from Gretley: mindful leadership and the law*, Sydney, CCH Australia Limited, 2007.

6   These issues are explored in much greater detail in Hopkins, A, *Safety, culture and risk*, Sydney, CCH Australia Limited, 2005.

Appendix 1

The use of "annualised" occupancy data for facility siting

The Amoco guidelines recommended the use of "annualised" occupancy data for facility siting risk assessments. This was criticised, both before and after the explosion.[1] The purpose of this appendix is to spell out the criticism as clearly as possible.

The starting point for a risk calculation in this context is the probability *per annum* of a vapour cloud explosion. When questions of blast pressure, trailer vulnerability, and so on, are taken into account, we end up with a probability of fatality *per annum* for a worker exposed to the risk of the vapour cloud explosion over a full year. Amoco's risk engineers noted, however, that if a maintenance shutdown is for three months, a trailer will only be at the exposed location for three months and workers will only be exposed to three months of risk rather than the full 12. Hence, the risk is only a quarter of the per annum risk.

The critics have argued that this disguises the true risk, making it "appear diluted when compared to the risk acceptability criteria".[2] Let us try to be more precise about this idea. The problem is this: dividing the risk by four because workers are exposed for only three months makes a critical assumption, namely, that the level of risk is constant throughout the year. If that assumption were correct, then it would follow that reducing the exposure time reduces the total risk to the worker. However, the risk of vapour cloud explosion is probably not constant throughout the year. Suppose we go to the other extreme and assume that vapour cloud explosions only occur during maintenance periods, perhaps because these periods involve startups and shutdowns which are known to be periods of heightened risk. In this case, workers on a three-month shutdown may be exposed to the entire per annum explosion risk in the three months of the shutdown. There would then be no justification for reducing the per annum risk to the worker just because he or she was exposed for three months rather than 12. The real situation is presumably not as extreme as this, but it seems likely that a disproportionate amount of the explosion risk is associated with maintenance periods. This is why the critics assert that the practice of reducing risk estimates, on the grounds that the exposure time is less than a year, results in an underestimate of the real risk.

---

1   *Investigation Report: Refinery Explosion and Fire*, US Chemical Safety and Hazard Investigation Board, March 2007, p 125; *Fatal Accident Investigation Report* (FAIR), London, BP, 9 December 2005, p 103.

2   FAIR, p 103.

Appendix 2
Process safety events

Figure 6.4 depicts a major accident triangle, developed by an Australian company, Santos. The company's principle process safety engineer presented its approach at an industry conference at Surfers Paradise, Queensland, in March 2008. The three lower levels of the diagram were further specified as follows:

Loss of containment
1. Large release.
2. Widespread release.
3. Release requiring site evacuation.

Process safety exception
1. Ignition sources created (eg fire, electrical, pyrophric, smoke, etc).
2. Non-compliance with applicable standards (Santos, Australian, International).
3. Permit to work breach.
4. Permit to work failure.
5. Unexpected extreme corrosion.
6. Valve left open (open port).
7. Changes without management of change.
8. Automatic detection/protection system de-activated.
9. Any crack, pinhole, etc, in process piping, leaking or not.

Process safety hazards and behaviours
1. Permit to work breach.
2. Incorrect tagging.
3. Inadequate blinds list.
4. Inadequate critical drawing.
5. Undesirable and frequent pressure safety valve operation.
6. Inaccurate or delayed response of a primary response system.
7. Changes without management of change.
8. Non-conformance to standards.

Appendix 3
The meaning of the arrows in Figure 15.1

Arrows in diagrams always need to be explained for a diagram to be fully meaningful. The arrows in Figure 15.1 have various meanings. Those that link the boxes in the accident sequence at the bottom of the diagram are necessary conditions, to be understood as follows: without the overfill, the release to air would not have occurred; without the release to air, ignition could not have occurred; without the proximity of the trailers, the deaths would not have occurred. This is a fairly precise chain of reasoning.

The four factors in the accident sequence have been explained using a single general concept: blindness to major risk. This serves to identify a crucial common characteristic. But there is a price to be paid by moving to this level of generality. The arrows no longer have such a clear interpretation. As noted in Chapter 13, in some respects the concept of blindness to major risk is not an explanation, but a description of what the four elements of the accident sequence have in common. From this point of view, the arrows leading down from this box might be read as "manifested itself in". Thus, the blindness to major risk manifested itself in the failure to replace the vent with a flare, and so on. In so far as blindness to major risk is an explanatory factor, then it is a necessary condition in the same sense as above. In other words, had Texas City not been blind to major risk, the operators would not have made the mistake they did, the vent would have been replaced with a flare, a vehicle would not have been left idling, and the trailers would not have been located where they were.

The arrow between "inability to learn" and "blindness to major risk" is a necessary condition: had the site been able to learn it would not have been blind to major risk. (It is not a sufficient condition because an intensive regime of inspections by the Occupational Safety and Health Administration might also have led to a greater focus on major risk.)

The boxes at the top of the diagram all refer to general ideas and it is hard to argue that each is a necessary condition, that is, it is hard to argue that, had any one of them been different, Texas City would not have suffered from an incapacity to learn. The arrows leading from these top boxes should therefore be interpreted as "contributed to" or, more precisely, "increased the likelihood of". Thus, the decentralised organisational structure increased the likelihood that Texas City would fail to learn from previous accidents, and so on.

This diagram therefore differs from those in my previous books. Those diagrams tried to maintain the logic of necessary conditions, or "but for" logic, throughout. That has not been attempted here because of the abstract nature of the concepts above the bottom line.

Index